SARAH'S SONG

Kim Smith

(Author of *The Dark Figure*, 1994)

and

Kevin Smith

with collaboration from

Dee Arianne Rockwood

PAGE PUBLISHING
Conneaut Lake, PA

First originally published by Page Publishing 2023

ISBN 979-8-88654-693-4 (pbk)
ISBN 979-8-88960-118-0 (hc)
ISBN 979-8-88654-700-9 (digital)

Printed in the United States of America

INTRODUCTION

There are some that believe the physical world is only part of our existence. Some believe there is also a spiritual world, which exists simultaneously and usually in balance with the physical world. Very few people have existed in both realms at the same time. An Indian shaman is believed to walk in duality with both the physical and spiritual worlds. This affords him the ability to foretell the future, heal the sick, and avoid detection in densely populated areas.

It is believed, in this realm, that what happens in the physical world also happens in the spiritual world and vice versa. Spirits can easily access both worlds. Physical beings, on the other hand, have great difficulty. Simply comprehending the possibility is exceedingly difficult. The physical world is plagued with limitations such as time and space. The spiritual world has few limitations; time and space are not among them.

If we were able to tap into this duality, we could travel in time while physically remaining in the present. This possibility has become a reality for several people. These people have traveled in spirit from one world to the next and back. To prove it to themselves, they commanded their spirits to bring back an insignificant item to place in their physical hand. Once accomplished, belief becomes knowledge, knowledge becomes reality, and reality becomes power—the power to believe in possibilities. These people wear those items around their necks as a humble reminder of where they have traveled.

Other religious beliefs hold this to be true in different forms such as religious prophets and apostles, those who have been enlightened by God or experienced and witnessed unexplained miracles of divine influence. These are examples of holy beliefs.

Witches and warlocks would be good examples of unholy beliefs and thought to be evil by some. Yet their power cannot be denied. People who can communicate with the spiritual world are said to be gifted. Mankind (in some cultures) was meant from the beginning to walk in the physical and spiritual worlds simultaneously.

These people who walk in this duality are considered strange in the eyes of the public. They are put away in institutions for being different. For some, hearing voices, seeing spirits, or following instructions provided by these spirits is a one-way ticket to the local insane asylum.

The ultimate battle between good and evil wages on. This war, throughout history, has maintained a delicate balance. The balance has drastically shifted in these modern times away from good. For thousands of years, evil has consistently gained ground. Man's insatiable appetite to possess more than he needs continues to upset this balance.

Mankind has enough bombs and missiles to annihilate all life on earth several times over. Still, it is not enough. Factories and laboratories are producing chemical toxins and biohazardous materials/agents at an alarming rate. Ironically, they are being used as weapons against people around the world.

If you were born after 1951, you have been exposed to radioactive fallout. That is now a fact. The scientific, military, and government communities have finally decided we now have the right to know, fifty years after being exposed. It is also a fact that hundreds of thousands of documents specify who, what, where, and how (we have been exposed) are not yet suitable for public disclosure. Maybe those documents will be released after we have died from complications of the earlier exposure.

Margin calls and capital returns are the buzzwords of profit-oriented businesses consistently investing in the destruction of our natural resources. Government leaders and agencies worldwide bark loud and often like a junkyard dog. Their bark has become old, tired, and useless. Rarely do they bite for fear they might lose funding for their campaigns or necessary support for projects that ultimately benefit them personally.

There is still hope among all this despair. An American archaeologist has uncovered the thirteenth scroll, which foretells of an imminent miracle, a miracle that is to take place in the twenty-first century, one that will have rejuvenating effects on all of mankind.

There were twelve disciples with Jesus Christ, and it is believed, even though they could not read or write, that they each had scribed their personal record of events. These twelve scrolls were not recovered by the time most consolidated versions of scriptures were published. And until today, they were still missing even with the discovery of the Dead Sea Scrolls.

In and of itself, the discovery of the views and events described in these twelve scrolls would be of phenomenal spiritual and historical value to the world. But even more exciting is the scripting of the thirteenth scroll, believed to have been authored by John the Baptist, the foreteller of Jesus Christ's coming.

The thirteenth scroll was never clearly documented as ever having existed until its accidental discovery by Tim Wilford, a museum curator. It is believed to be written by John the Baptist shortly before he was beheaded at the whim of a jealous queen. Its pages record a miracle that is to occur sometime in the twenty-first century that will change the balance of spiritual power and bring peaceful harmony to the world.

Time is of the essence for Tim Wilford and his partner Haseeb Jaffar. They must get the scroll's contents published worldwide for it to have the intended effect. The last miracle is about to take place, and no one knows about it. Sometimes man himself plays the most important part of his own awakening.

CHAPTER 1

Protecting the Thirteenth Scroll

Early May 1950, at an undisclosed Tibetan Buddhist temple in the Tibetan Plateau, a letter was received. Written in Mandarin Chinese, the letter, loosely translated, warned of impending danger to a temple in South Korea near Suwon. A sacred scroll of inconceivable religious value needed to be safeguarded at all costs. The Korean temple could not guarantee the safety of the scroll any longer because of current political agitation. They beseeched the Tibetan temple to assume responsibility for the priceless parchment. Enclosed in the letter was a golden key.

The elders met to decide whom from their temple would be best suited for this mission. Little time was spent in the small stone room before a name was chosen. Myung-Dae Kwon came to the temple five years ago after losing his wife and child in a horrible fire that destroyed his home and his life.

Since his arrival, Myung-Dae Kwon, a thirty-one-year-old South Korean farmer, had demonstrated his loyalty, faith, and reliability. Moreover, he was still connected to the outside world and was remarkably familiar with the Korean territory. His five-year stay at the temple had not been so long, and the outside world would pose little challenge for him.

The decision was made, and Myung-Dae Kwon would leave at the end of the week. Tao, the sixty-five-year-old elder and leader appointed by the Dalai Lama himself, blessed Myung-Dae Kwon before sending him on his journey. Tao was orphaned at age two and left at the Tibetan temple in November 1885. He took to the monks' way of life, embracing it, and never had a reason to leave.

It was a warm and humid morning in late June of 1950. Myung-Dae Kwon and his faithful boatman Kim poled their longboat to the banks of the Han River just east of Suwon, South Korea. The sun just peeked through the tall, green palm trees, casting long, dark shadows over the plush jungle landscape. Myung-Dae Kwon, clothed in a beige robe, jumped from the longboat to the river's bank and ran barefoot up the snaking, muddy path toward the smoldering remains of the temple. With unrelenting determination, he quickly scurried through the rubble.

Once, a beautiful, majestic Buddhist temple stood here, protected on the east, south, and north by miles of treacherous jungle and accessible only by the most skilled navigators of the treacherous Han River. The temple was considered well-insulated from political turmoil and a sanctuary to all in search of peace, tranquility, and spiritual harmony, now obliterated by man's unquenchable thirst for power, fueled only by greed, and set in motion by his insatiable appetite for control during political unrest.

Myung-Dae Kwon was about 5'6" tall, 145 pounds with defined muscles and a slight pooch belly. He had short black hair and brown eyes that seemed to hold many secrets. On this day, his face was chiseled with grief as he knelt and brushed the warm ashes to one side, revealing the cast-iron floor safe. Nervously he glanced around, surveying the temple grounds and surrounding jungle for any signs of intruders. The varying thickness of white and gray smoke, rising from the smoldering fires that littered the landscape, made it difficult to see clearly. Reaching into his robe, he removed a large gold key and placed it into the keyhole.

Myung-Dae Kwon clasped his hands together in prayer and bowed. Completing his prayer, he turned the key and opened the safe. Turning his head to one side, he reached shoulder deep into the opening. After fumbling round for a few seconds, Myung-Dae Kwon carefully removed the sacred scroll wrapped in goat hide. He handled it with deep reverence and respect. He unwrapped the precious document from the goat's hide and delicately and meticulously examined it to ensure it was in one piece. Now satisfied that it was undamaged and safely in his possession, he sighed with relief.

Myung-Dae Kwon glanced over the surrounding terrain cautiously, remembering the critical nature of his mission. Now that he was in possession of the sacred scroll, his life was in great danger. He momentarily recalled the instructions given by Elder Tao: "Against all peril, you must bring the scroll safely to Tibet."

He carefully placed the scroll into a leather case under his garment. With great stealth and speed, he departed the temple ruins. He maneuvered swiftly down the muddy, snaking path toward the river, where Kim his boatman should be patiently waiting for him. A hundred yards down the path, his senses tingled, a sign that there was danger lurking.

Forming directly in front of him was a huge rift in the air, a large vertical, oval mass surrounded by dark black clouds. The oval shape consisted of a transparent substance that resembled rippling water and hovered a foot or so off the ground. Its diameter was approximately twelve feet, and from this portal emerged a large, hideous creature of immense proportions.

He stood ten feet tall from foot to horn. His muscular, outstretched neck supported a jackal's head with two ram horns, one on each side. His ominously dark human skeletal shape was covered with a tarry substance that was black as midnight. He stared at Myung-Dae Kwon through two glowing crimson eyes.

The creature let out a bloodcurdling roar that could be felt in Myung-Dae Kwon's bones as he stepped forward, clearing the rippling portal. His feet consisted of three toes, each hosting a ten-inch razor-sharp talon and a single fifteen-inch claw on the back of his heal. His sharply defined, muscular arms were exaggerated in length, extending seven feet from his shoulders, which stretched out toward Myung-Dae Kwon. The creature's hands had four fingers with large five-inch claws at the end of each and an inverted thumb with no claw.

In one hand he carefully cradled a large, elongated, dark wooden box. In the other hand, he held the lid. A long, spiny tail protruded from the end of his spinal column with an arrowhead tip at the end of it. Myung-Dae Kwon was petrified with fear.

As the grotesque figure stepped closer to Myung-Dae Kwon, his stomach convulsed from the foul stench that enveloped him. The servant of Satan planted himself firmly only a few feet from Myung-Dae Kwon and bellowed with a voice that shook the ground for miles. His putrid breath dropped birds in flight and frogs from the trees to the ground. It laid the tall wheat grass down around the path as far as one could see.

At the bank of the Han River, Kim heard the unholy roar of the beast and fell back into the longboat. Peering over the side, he saw the river water dancing (as it does when an alligator sounds its mating call). Kim saw the birds falling from the sky and the frogs dropping into the river and along the shore. He was struck with fear.

The repugnant creature spoke to Myung-Dae Kwon with a horrifying voice: "In your cloak you carry a script that belongs to me. Put it in this box, and I will not harm you." Myung-Dae Kwon not only heard the words from the beast but he also felt them, echoing in the marrow of his bones. Satan's spawn laid down the open box at the monk's feet and stepped back, expecting unquestioned compliance.

The hellish creature waited patiently for the monk to decide his own fate. Myung-Dae Kwon did indeed contemplate but not his own fate; he reflected on the faith that the elders of the temple had placed in him. He thought why this great, powerful servant of Satan was asking him to place the scroll into the box. He wisely surmised that the creature was incapable of taking the scroll from him.

After several moments, which seemed like hours, he defiantly refused to comply with the mighty creature's request. He would not yield to the horrifying bantering of this hellish beast. Instead, he would honor those who sent him on this mission by denying the beast's request, knowing full well that if he was wrong, he would be condemning himself to a horrible fate.

He looked up into the towering creature's eyes and said, "I, Myung-Dae Kwon of the temple of elders in Tibet, deny you and your masters' request for this sacred scroll. It belongs to the righteous servants of the ancient one and shall not pass from my hands to the likes of you. Now be off, you servant of chaos! Return from whence you came, and remove yourself from my path."

Myung-Dae Kwon slowly sat down on the muddy path and assumed the lotus position and began to chant his mantra quietly, calling for guidance from the council of elders. He closed his eyes and put the creature out of his mind, but only for a moment; the beast had great powers, and soon he forced Myung-Dae Kwon to open his eyes and face him.

"You fool!" he roared in a thunderous voice, shaking the landscape again. "I have no time for your petty insolence! Place the scroll into the box, or I will tear you into a thousand pieces," the creature coarsely demanded.

Myung-Dae Kwon stood up and planted his feet firmly. He was certain that the creature could not harm him as long as he was in possession of the sacred document. But he was a little nervous because this creature was persistent to a fault. He suspected that the battle was only just beginning and was concerned that the demon may try to trick him.

His suspicion was ratified when the creature spoke. "Surely there is something I could trade you for that worthless item. After all, the document has no value to you—it is only important to those called Christians." Myung-Dae Kwon saw a hint of wisdom in the monster's words. "Name your price, monk, and I will give it to you," the creature bellowed with confidence in a non-threatening voice. His tone had changed completely, and now the low hum of his voice rolled over the monk's ear like the perfect mantra, soothingly relaxing Myung-Dae Kwon's senses. An image of the elders' counsel cleared the spell, and Myung-Dae Kwon was able to shake off the trance the monster had invoked.

"There is nothing you can offer me. Now stop with your tricks, you siren of suffering, and stand aside," Myung-Dae Kwon demanded with conviction and confidence.

The monster snorted like a bull and continued with his own confidence as if he were acutely aware of the monk's price. "Are you so certain there is nothing I can offer you?"

There was a long pause between the dialogue as they stood eye to eye, neither budging. Suddenly, images of Myung-Dae Kwon and his family flooded his thoughts and danced in his head. He saw himself playing with his son in the park near their humble home in Taegu. His wife was preparing the picnic lunch just before heading to the park. Myung-Dae Kwon collapsed onto the ground in grief, sobbing uncontrollably over the horrible loss of his family.

The demon smiled as if victory were placed in his hand. "What if I could return your wife and child to you with enough gold and jewels to live like a king? You could spend the rest of your natural life together in happiness and peace. Certainly, that would be a fair trade for a piece of old parchment that means nothing to you."

Myung-Dae Kwon lay prone on the ground, unable to stop weeping over his loss. The creature grinned from horn to horn, exposing all his yellowing canine teeth. "Look here, my friend!" He pointed to his right side, and standing in the dead field next to the beast were Myung-Dae Kwon's wife and child, swarmed by impish servants. "As I promised, here is your wife and child."

"Look over here!" The hideous creature pointed to his left side, and the dead field was filled with shining bricks of gold and pots of sparkling jewels. "Here is the gold I promised and the jewels to adorn your beautiful wife." Myung-Dae Kwon stopped weeping and focused only on his lovely wife and child.

"Now put that old, worn-out script into the box, and take all these gifts with my blessing," the demon urged gently. Myung-Dae Kwon reached slowly into his robe and withdrew the leather pouch and removed it from his neck. The beast was anxious for his victory and kicked the wooden box closer to Myung-Dae Kwon.

Myung-Dae Kwon wrapped a portion of the thin leather strap around his right index finger and held the canister in both hands in front of him. Hiding his finger from the creature's view, he placed the document into the wooden box. He bowed his head, letting the shiny coin medallion his wife had given him lay flat on the ground in front of him.

Staring at the monster's reflection in the medallion, Myung-Dae Kwon watched the beast very carefully. The beast slowly raised his right hand high above his head to deliver a lethal blow to the stupid monk. As the demon started his swing, Myung-Dae Kwon jumped to his feet, pulling the holy scroll out of the wooden box and holding it up to block the creature's deadly blow.

Before the servant of darkness realized the monk had tricked him, he struck the scroll with all his might. The blow ignited a white spark that shot into the heart of the beast and knocked Myung-Dae Kwon to the ground. The creature grabbed his chest and screeched in pain as his body disintegrated into a thousand pieces.

The images of Myung-Dae Kwon's wife and child disappeared along with the gold and jewels. The impish servants exploded into pieces that started circling above the beast. All the pieces of the monster and impish creatures twisted into a large crimson vortex and were consumed by the portal. In the blink of an eye, the portal disappeared.

The scars in the earth from this great mayhem remained as a permanent reminder of the power of evil. The birds that fell from the sky and the frogs from the trees remained lifeless on the ground. Sadly, there was nothing that Myung-Dae Kwon could do for them. As he sat up, Myung-Dae Kwon reflected on how close he came to becoming a sad part of history. He heard a familiar sound from the path below.

It was Kim, his boatman, scrambling up the path to see what had happened. Kim approached cautiously, slowing his run to a walk as he scanned the field of battle. His eyes surveyed the dead fowl and reptiles scattered throughout the tall grass flattened down like a green carpet. "Myung-Dae Kwon! Are you well enough to travel? This place smells of death."

He walked over to where Myung-Dae Kwon was still sitting and helped him to his feet, brushing the blades of grass off his robe. "What evil has caused such destruction in this beautiful place?"

Myung-Dae Kwon rose to his feet. "Exactly what evil I cannot say, but that it was truly evil I can affirm to you. Come! We cannot delay. We must get back to the temple with haste."

Myung-Dae Kwon placed the leather case under his robe and secured it tightly around his neck. They swiftly walked down the trail to the long, narrow wooden boat that sat on the shoreline of the Han River. Kim asked him, "Was it the scroll that caused the evil one to attack you?"

Myung-Dae Kwon looked at Kim sternly and replied, "You can tell no one of this, my friend. It would put fear in the hearts of brave men and cause women and children to weep. Tell no one!" Kim assured him he would remain silent about the event and what he saw. They boarded the launch and started back up the river.

CHAPTER 2

The Monks' Return

The journey back to the Tibetan temple was a long and dangerous one. Kim navigated the Han River north and east to the Yellow Sea. There they rented a larger boat to navigate east to Shanghai, towing their long boat behind. Neither spoke a word during most of the journey. An occasional glance or nod to confirm the direction of travel was all that was required.

In Shanghai, Kim prepared his long boat for the journey through China. Civil unrest, deep in the central mountain regions, would make it necessary to remain on the craft for several days at a time. Kim modified his boat with a small platform on one side to hold their meager rations and a pontoon to stabilize and counterbalance the extra weight on the other.

Kim and Myung-Dae continued their journey traveling east on the winding Yangtze River through the center of China to its end just north of the Tibetan Plateau. When they arrived at the shore, this was where Myung-Dae Kwon's journey would be continued on foot. Myung-Dae embraced Kim and offered him a blessing for his service. He pulled out his purse to pay him for his company and the use of his boat. But Kim was happy he could serve and accepted the embrace as full payment, taking nothing from Myung-Dae Kwon.

Myung-Dae traveled south through deep valleys and tall mountain ranges. It was exhausting, even for a man of his relative youth and stature. The trip had seemed to age him beyond his years. His feet were as heavy as ten-pound lead weights; his leg muscles burned and ached with each step as he progressed. Three grueling months and 3,600 miles after his departure, he finally returned to the beautiful hidden sanctuary of his Tibetan temple resting high in the mountains south of the Tibetan Plateau.

Myung-Dae Kwon was greeted at the temple entrance by all the elders led by Tao. Tao was most pleased to see Myung-Dae Kwon. Tao greeted him with a hug and a smile. "You have done well, my son. The spirits are at peace for a time." Myung-Dae Kwon started to explain what had happened at the old temple and his encounter with the demon.

Tao and the group of elders stopped as Myung-Dae Kwon continued to walk and talk about his adventure. Myung-Dae turned to see what might be wrong with Tao and the other elders. They were gathered talking among themselves when Tao stepped out from the group, looking at Myung-Dae Kwon, and said, "We are aware of your adventure, Mr. Kwon. We saw it very clearly." There was a long pause as the elders looked deep into his eyes. "We were with you, my friend." Tao extended his right hand slightly with the palm of his hand facing skyward and turned to the group slowly. As he pivoted toward the elders, they bowed from the waist politely, then Tao faced Myung-Dae Kwon, and he bowed.

Myung-Dae Kwon finally realized he had never been alone. The elders had traveled with him in spirit and were there to strengthen him when he was confronted by the demon. Myung-Dae Kwon fought back the tears as he bowed to them humbly in appreciation for their assistance and in respect of their choice to send him.

Tao turned back and put his hand on Myung-Dae Kwon's shoulder and escorted him toward the temple. "We were like small birds in a tree, always with our gaze upon you. We witnessed your entire journey. You displayed much honor, and we are incredibly pleased. Now go and rest. It's been a long trip. Your body is tired."

He heeded Tao's wishes and admired the old man's wisdom. He retired to his room and lay down on his bed. Before a moment had passed, he fell into a deep sleep. He slept for a day and a half. The elders watched over him.

When he awoke, Myung-Dae Kwon was starving and headed straight to the kitchen, where a meal had been prepared for him.

The elders summoned Tao once Myung-Dae Kwon had woken. Tao needed to speak with him right away. Tao met with Myung-Dae Kwon at a table in the open temple courtyard just outside the kitchen entrance.

"I see you are well-rested and have resumed your appetite," Tao exclaimed as he entered the courtyard.

"Yes, Master, it is my appetite that woke me."

Tao patted Myung-Dae Kwon on the shoulder as he sat down next to him. "Good, good, I am happy to hear that." Tao smiled for a moment as Myung-Dae Kwon continued to consume his meal. "We must talk about your encounter with the evil one." Myung-Dae Kwon looked into Tao's eyes and saw that he was concerned with what he had to say. Tao's smile had faded.

Myung-Dae Kwon sat up and pushed the bowl of grub to the center of the table and turned to give his full attention to his master. "Yes, Master, please! Say what is on your mind. It is well with me that you wish to speak of this." Myung-Dae Kwon pulled the bowl back and continued to eat, wanting to ease Tao's hesitation to speak.

"Battling demons is usually reserved for more enlightened holy men, those who spend their entire lives devoted to preparing for one battle. You were chosen because of the purity of your heart and soul, not because of your formal preparations."

Tao continued, "Your battle was a great one against the evilest of all demons. Rochoa, son of Satan, he has never failed in battle until he met you. The reason I tell you this is to prepare you for your next journey. A journey to the other side."

Myung-Dae Kwon pushed his bowl back to the center of the table again. He looked intensely into Tao's eyes and inquired, "Another journey, Master? Where will I be sent now, and when will I go?"

Tao's eyes began to well up with tears as he explained to Myung-Dae Kwon, "Your next journey is beyond your physical body. I cannot tell you when, but it will be soon. You will no longer be constrained by the limitations of the physical world. Your new journey will move you into the spirit world, and you will no longer have need for this shell." Tao tapped Myung-Dae Kwon on the shoulder, indicating his death.

Myung-Dae Kwon looked surprised and puzzled at Tao. "You mean I am to die, Master?"

Tao nodded. "Yes, but you must understand that death is only the end of one great journey, not just the passing of a life. Your mission continues in another world where your spirit is so desperately needed. You will now be our guide as we have been yours. It is a great honor for you, and I wanted to wish you well before you start."

He stood, as did Tao. The master's words were very sobering. Myung-Dae Kwon smiled at the old man and embraced him firmly. "I am honored. You and the elders have been a great inspiration and a source of strength for me. I only hope that I may be equal to the task."

Tao turned and walked out of the courtyard, shaking his finger at the sky. "Oh, you will be, my friend, very much so!" He continued out of the courtyard to meet with the elders and prepare them.

Myung-Dae Kwon sat back down and let his mind flood with memories of his beautiful wife, Eun Ae (which means "grace with love") and his son, Beom-Seok (which means "like a rock"). Perhaps now, he hoped, he would be reunited with his family.

He recalled how happy they were together carving out a meager life for themselves on the farm his father and mother had built in a valley just west of Tague, South Korea, with high mountains surrounding three sides to the south, east, and west. The Nakdong River was to the north. The work was hard but extremely rewarding, knowing he had the means to care for his family.

He reminisced how much they enjoyed swimming and fishing in the river and spent time together hiking into the mountains that surrounded their farm. Myung-Dae Kwon wondered if they would be able to do these things in the spirit world. But for now, he wondered if crossing over to the spirit world would be painful or if it would be as simple as going to sleep.

He went into the temple prayer chamber and collapsed to his knees for a moment then sat on the floor in front of the altar, where a large Buddha statue clad in gold leaf sat looking over the spacious room. Myung-Dae Kwon crossed his legs with his feet resting on his knees and laid his hands on top of his feet with the palms facing skyward. Then he began a mantra, humming tunes softly to start his meditation and prepare himself for his new journey.

Over the next several weeks, Myung-Dae Kwon settled into a regimen of eating two meals a day and spending the rest of his time in prayer and meditation. He never remarked to anyone what Tao had shared with him. Three weeks after his return to the temple, Myung-Dae passed away quietly in his sleep. He was at peace knowing that his mission was successful and the sacred scroll was safe on holy ground.

The thirteenth scroll rested safely in a ground vault under the gold-leafed Buddha. It had been placed there by Tao shortly after Myung-Dae Kwon had returned from his journey. Behind the Buddha statue was a secret door that could only be revealed by touching secret places (known only to Tao) on the front of the statue in a specific order. Once opened, Tao descended the steep steps down to a dark room and lit a torch to shed light on the ground safe. Tao took a gold key hanging from a chain around his neck and inserted it into the lock. Then he lifted the cover off and placed the scroll into its new resting place. He secured the cover, locked it, put out the flaming torch, and ascended the stairs, exiting the door behind the Buddha. Tao walked back to the front of the statue and in reverse order touched the secret places to close the vault door.

The thirteenth scroll remained there, undisturbed for twenty-one years, until one warm day in June 1972, a stranger arrived. The remote Tibetan temple was not in the habit of receiving unannounced visitors and quickly protested the intrusion. Tao moved forward through the swarming crowd of monks and witnessed the stranger. He then raised his hand to silence them, and they complied. Tao informed his brothers, "This man is my guest. I have been expecting him for some time now. Please go about your business and studies so we can visit in peace." The elders and monks disassembled and went back to their normal routines.

The visitor's name was John. He was about six feet tall, of slender face and build. He had shoulder length auburn hair and strikingly blue eyes. He was dressed in a light brown-colored pullover, which looked like camel hair. He wore matching loose-fit slacks that were cinched at the waist with a hemp cord. His feet were wrapped in a pair of old, dilapidated leather sandals that were tethered above his ankles. His skin was golden brown and sparkled in the sun as though he had been sprinkled with gold dust.

Tao knew why John was there. He had been anticipating his arrival for twenty-one years. John was one of the truths foretold in a state of heightened awareness during his many meditations. Tao was a great sage and had experienced many visions during meditation. One that brought him great joy was envisioning Myung-Dae Kwon passing over in peace and being reunited with his family after completing the task presented to him. But this one about John was the most important of all.

On the eve of Myung-Dae Kwon's death, Tao had been chanting his mantra for several hours when he was told by an angel, "A stranger called John will come to visit you at the Tibetan temple during the summer of 1972. You will know him by his virtuous heart. You will not taste the sting of death until this comes to pass. You will present him with the sacred scroll placed in your possession. Then you will be called forward to join the elders that have passed before you."

The visitor, John, stayed with them for a few days. He blessed the temple and those that lived within its walls. John taught the monks and elders how to heighten their awareness and lead them to a more enlightened meditation. He spent many hours alone with Tao in the temple garden, imparting great wisdom and truth.

On the third day of his visit, John informed Tao that it was time for him to depart. Tao retrieved the thirteenth scroll from its hiding place and presented it to him and bowed. John turned and left, walking out through the temple entrance; the monks and elders showered his path with rose petals and bowed as he passed them. As John reached the boundary of the temple's holy ground, he turned back slowly and faced the temple, where he offered a final blessing. He then turned his back to the temple and moved swiftly, floating over the terrain for a few seconds, and then vanished into thin air.

John reappeared inside a buried temple on the outskirts of Cairo, Egypt, a temple that had not been discovered yet. He moved through the passageways of the temple to an open room, with two guardian sarcophaguses placed in the center. He walked behind a stone altar, where a stone shelf displayed twelve similar-sized scrolls. He gently placed the thirteenth scroll beside the rest and blessed them. John knew that soon, an archaeologist would discover this temple and the secrets it contained. He would realize the great significance of the thirteenth scroll and attempt to release it to the world. Few men had been privileged to view the contents of this scroll. None had lived long enough to realize its great value. With his mission completed, John vanished again.

On this same day back in the Tibetan temple, Tao sequestered himself to isolation in the sacred room. He silently digested the great wisdom and knowledge John had revealed to him in the garden. Fifteen days later, he assembled the monks and elders together and shared with them the wisdom that he was allowed. Two days following this event, Tao passed peacefully in his sleep just as Myung-Dae Kwon had.

CHAPTER 3

Heartbreak for Evans

It was September 1980 at Westwood High School in Mesa, Arizona, where Mike Victor Evans, a pimple-faced, slender-built but physically well-defined teenager, was captured slamming his locker door and running down the hall. Mike was about 5'10" tall, weighed about 145 pounds (soaking wet), with brown hair and brown eyes, and sporting a bronze tan as most active students did in the desert valley. Mike was a junior here and was looking forward to getting to his car and beat some of the high school traffic.

Mike had no idea of the critical role he would soon play in bringing forth the great miracle described in the thirteenth scroll. He was also unaware that the girl he was about to meet would play an even bigger part in this event. Right now, he needed to get home to change his clothes and meet his friend Bernard Scott. Bernard and Mike had been good friends since they met a year ago when Mike first started at Westwood High School. Bernard had convinced Mike to join him at St. Timothy's after school for the Bible study group. Mike agreed because the theological opinions of others intrigued him and opened him up to other windows of study.

As Bernard and Mike arrived at the Bible study group, everyone was there and scuffling to find seats. Mary Ann Brinkley was already seated at a table and looking through scriptures. Mary had been in the study group for nearly a year now and was familiar with the routine. She was a very conservative teenager, relying on her academic abilities to pave her way to college on a scholarship. Mary had beautiful long, brown hair (reaching down to her butt), big, brown eyes, high cheekbones, and supple lips. Mary was blessed with an amazing figure and long, well-shaped legs. Her skin was tanned evenly with a bronze appearance, and when Mary smiled, she made everyone feel comfortable.

Bernard announced to everyone, "Hello, everyone. I have brought a guest with me today. His name is Mike Evans, and he is a good friend of mine. Please make him feel welcome." Mike was a charming, intelligent teenager who had a charming smile that warmed the hearts of those he met, including Mary. Mike scanned the people in the room, acknowledging their friendly greetings until he saw Mary. His eyes locked onto Mary for several pulse-pounding seconds. Mary too was unable to disengage her stare into Mike's eyes, and both became spellbound. After several seconds, the trance was interrupted by Bernard's voice. "Are you okay, Mike?"

Mike turned his gaze back to the rest of the group, who had appeared to notice the exchange, and coughed as though he had a cold. "I am fine. I just have a little cold, and the medication I'm on sometimes causes me to drift off occasionally. But really, I am fine. Hello, everyone, it is nice to meet you." Mary looked back down at her scriptures with a cheeky smile. Mike quickly sat down in an open seat across the room from Mary.

For those young and impressionable teenagers, those few seconds set the tide of love in motion. In the case of Mike and Mary, it was a tsunami. Little did they know they were destined to be together for the rest of their lives. Throughout the rest of the Bible study group, they could not keep their eyes off each other. When the meeting concluded, Mike was able to summon the courage to walk over and speak with Mary. But first, he had to conjure up images of traffic accidents and gutting a deer to get rid of his boner. After all, he was still a typical teenager.

Below the surface, in the intellectual portion of Mike's brain, he was going to ask, "*Mary, I could not help noticing a magnetic attraction between us. Would you like to join me for a soda later so we can discuss this strange and wonderful occurrence?*" However, what took place was much different. "Murray...coke...meee...k?"

Impervious to Mike and Mary, the room around them was filled with joyful Christians holding their stomachs, gleefully trying to suppress uncontrollable bouts of laughter. Mary simply pulled her handkerchief out of her purse, delicately dabbed the drool pouring from the corner of Mike's mouth, and accepted his request with a smile and a nod. They turned together and left the room. The laughter faded away with more distance.

From that moment on, their romance blossomed into a beautiful courtship of emotions. Of course, to catch a glimpse of any silver lining, you must first peer through dark clouds. Mike and Mary sure had their fair share of emotional and social thunderstorms, most of which stemmed from their distinctively different backgrounds and social circles.

Mary came from a devout Roman Catholic family. Their faithful attendance at mass and other church functions was laudatory. Mike was raised in the Presbyterian faith, attending church on occasions such as Easter Sunday and Christmas. Reading the Bible on Sundays was a particularly important practice instilled in him while he was young.

Mary came from a split home. Her mother, Agnes Brinkley, lived in South Phoenix with Mary's brother David and sister Susan. Mary had moved from South Phoenix to Mesa, Arizona, in March 1981 with her father, Paul Brinkley, his new wife, Gloria, and her stepbrother Bob and stepsister Loraine. She had made the move that year because things were getting serious between her Mike, and she wanted to be closer to him. Mike lived with his family on Horne in Mesa, Arizona.

Mike was an Air Force brat and had traveled the world from Japan to Guam, Hawaii, Germany, England, Illinois, Alabama, Florida, Washington, DC, and New York before he was sixteen and the family settled in Mesa, Arizona. His father, Frank Evans, was a retired Air Force pilot who was wounded in Vietnam when the North Viet Cong detonated an explosive device smuggled into a popular off-base restaurant near De Nang Air Base in the summer of 1963. His back was broken in three places, and he could not fly any more. After being grounded and placed in the War Plans Department of the Pentagon, Lieutenant Colonel Frank Evans retired and accepted a position with Link, Singer division in Vestal New York. After a few years, he accepted a new civilian position with the Air Force as chief phycologist of the Flight Simulator Division at Williams Air Force Base in Chandler, Arizona.

Mike had two brothers. Robert was nine years older and Dick his identical twin brother. Robert was married and living in their old house in Tampa, Florida, on Bay Court Avenue with his family. Dick was secretly preparing to sail off into the United States Navy because he just could not handle the boredom of Westwood High School.

Mike's mother, Virginia, was the Norman Rockwell picture-perfect mom. She took such good care of "the boys," as she liked to call them, that she probably spoiled them. She was always active in their lives and seemed to live for the moments that she could help them out. Virginia was always cheerful and giving in nature; she spent a great deal of her free time with the ladies' auxiliaries and charitable organizations.

Mary's mother was a wonderful spirit. She was strict with the kids but also very protective and loving. Agnes was one of nine children, and the family blanketed the Phoenix valley. The Brinkley family reunions were held outdoors in large parks, and attendance numbered in the hundreds depending on the occasion.

On both sides of the families, there were those members who tried to encourage Mary and Mike not to see so much of each other. "There are a lot of fish in the sea. You should catch a few more before bringing one into the boat," both families would say to discourage the two from becoming any more committed. Mary and Mike remained polite and receptive to their comments but never gave them much thought; they were in love, and that was all that mattered to them.

As time passed, the well-intentioned family members and friends gave in to the delightful couple's charm and finally accepted the inevitable. Mike got a job as an electrician while still attending his senior year at Westwood High School with Smith Brothers' Construction, working through the summer to save money for the future.

After graduation from Westwood High, Mike became a journeyman electrician, doubling his hourly wage. Mary still had one more year left before she would graduate from Mesa High. They say that was the longest year of their lives. Mike and Mary had promised that they would not move further in their relationship until they had both finished high school. Mary graduated Mesa High in June 1982. In August of 1982, Mike proposed to Mary with her father's permission, and she gladly accepted.

Finally, on June 5, 1984, they were married at Our Lady of Mount Carmel in Tempe, Arizona. It was a beautiful wedding with all their friends and family gathered to celebrate the happy occasion. After the ceremony and pictures, everyone was invited to the activity center for cake and punch. Later, they moved the celebration to the Evans home, where the punch was a little stronger.

It was there that Frank Evans handed the couple the keys to a new car as a wedding present and their reservation for the Honeymoon Suite at the Double Tree Inn in Scottsdale, Arizona. After a few hours of being polite with the guests, Mike and Mary slipped out the kitchen door to the carport, jumped into their new car, and raced over to the inn. They checked into the hotel under the name of Mr. and Mrs. Mike Evans. The suite was beautiful and was filled with flowers and a bottle of champagne on ice in a silver bucket with two glasses.

They stripped naked after bolting the suite door, grabbed the champagne, popped the cork, filled the two glasses, and headed to the bathroom. There they took a nice, relaxing shower together while finishing their champagne. They dried off and went back into the center room where there was a nice couch and coffee table. They had two more glasses of champagne while admiring each other. Mike signaled it was time to get into bed without having to say a word (he was fully erect). Mary quickly jumped into the bed and stretched her arms out toward him, signaling for an embrace. They turned the lights out and held each other close and made steamy, hot, passionate love. They finally fell asleep in each other's arms.

The next morning, they went down to the first floor where the restaurant was and ordered breakfast. While they were waiting for their breakfast to be served, Mike looked into Mary's eyes and expressed his love for her and assured her that he would take good care of her. Mary smiled and said, "Right back at you, cupcake!" They finished breakfast and went to check out.

Mike learned the construction business quickly and applied his new skills to their own home. He was promoted to site foreman on one of the biggest projects Smith Brothers' Construction had in the area. Bill Smith, owner of the Construction Company, kept his eye on Mike; he was extremely impressed with his enthusiasm and hard work habits. Mike ran his own crew as a working foreman and loved it.

Mary was working for Bob's Big Boy Restaurant on Main Street in Mesa. The hours were tough, but the tips were good, and she loved the work. Mary did not really like closing the place at night; it made it hard for her and Mike to spend the kind of time together that they wanted (they wanted to spend all their time together). She put a few applications out for a banking position in the area. After a few interviews, she was hired as a bank teller at First National Bank near their home in Mesa. Mike and Mary had purchased a small three-bedroom, two-bath home on Staple Street in anticipation of children (someday).

Several years passed as they fine-tuned their relationship and established their daily routines. Mike continued to grow and excel as a leader with the company. Both Smith brothers (Bill and Fred) stumbled over themselves trying to land bigger and better contracts for Mike to oversee. He had made a real name for himself as being someone that got the projects done early and under budget. Mary was doing very well for herself with the bank. She was promoted to branch manager within two years and in line for a vice president position.

Every evening at dinner, they discussed family matters and went over their finances. Financially, they were looking good. They had managed to save over twenty-five thousand dollars in the past couple of years. They talked about having a child to complete the family. They wanted to plan their family's growth carefully to ensure they were able to meet the financial and nurturing needs of a baby. But as we all know, even the best-laid plans backfire.

On a warm day in March of 1987, Mary was in severe pain and called Mike at work. He bolted from the job site and drove straight to Dr. Robert Kissling's office to meet Mary. Dr. Kissling sadly informed Mary, with Mike at her side, that she had ovarian cancer and they would need to operate right away. The doctor explained that he would not be able to provide any further information until he got in there and checked everything out. Mary was checked into the Desert Samaritan Hospital that day.

Mary was put through a battery of tests and was scheduled for surgery the next day. Mike called work to notify her boss and then call his and her parents and asked them to only let immediate family know of her condition right now. He did not want a crowd raining down on them at the hospital when they were not sure what the damage was. They all agreed. Mike then called Mary's boss to let her know what was going on and that she would not be into work for a while.

The next morning, Mary was prepped and rushed into surgery. Mike, his father, Frank, and his mother, Virginia, along with Mary's parents, Paul and Agnes, paced nervously in the waiting room as other family members started arriving. Before long, the waiting room was full of anxious family members waiting to hear any news of Mary's condition. Mike, Frank, and Mary's dad, Paul, moved their nervous parade to the hallway.

Finally, four long hours after Mary had gone into surgery, Dr. Kissling came out and approached Mike. He grabbed him gently by the arm and escorted him into a vacant office across the hall from the waiting room and explained, "Mike, Mary suffered a serious infection of her ovaries. The cancer had pretty much destroyed her childbearing organs. I am sorry, but we had to perform an emergency hysterectomy. If we did not take immediate steps now, the cancer would have spread throughout her body and the prognosis would have been much worse."

Mikes eyes immediately swelled with tears, and he broke down for a moment, crying uncontrollably, then wrestled to pull himself together. "Will Mary be okay?"

Dr. Kissling looked into Mike's red eyes and forced a confident grin. "Yes, Mary is going to be fine. She will not be able to have children of her own. But you both can still raise a beautiful family together through adoption. We were able to catch the cancer before it spread to any other organs. I expect you will be able to take her home in a few days."

Mike was glad to hear that Mary was going to be fine. But it was bittersweet because he knew how much Mary wanted to have children of her own. "When can I see her, Doctor? Does she know yet?"

Dr. Kissling walked over to the door to leave. "You can see her in recovery, she should be down there in about an hour, and no, I have not told her yet, but I will before you see her. Mike! I am so sorry—we did everything we could." Then he left the room and walked past the waiting room into the double door of the operating rooms.

After composing himself, Mike exited the small room and went into the waiting room to talk with the rest of the family. As Mike explained what the doctor had told him, he was overcome again with despair and broke down. The family comforted him and prayed for both.

When Mike went into the recovery room to see Mary, she was awake and wiping her eyes with tissues that were bundled up in her lap. Mike walked over and kissed her on the forehead then embraced her gently. He whispered into her ear, "Everything will be fine, honey! God has a special family for us, and he will let us know when. Of this I am sure. I love you!" And he kissed her again. With that, Mary started to pull herself together for him.

Just then, there was a slight cool breeze that filled the room. Hovering at the foot of the bed was the traveler John, who had recovered the thirteenth scroll from the Tibetan Temple and placed it with the other scrolls at the hidden temple in Egypt. He was still wearing the same robe and sandals and showed no aging after all these years. He gazed caringly at the couple with his hand raised to the sky as though he was blessing them or protecting them. Mike and Mary felt the breeze but could not see the image that had appeared.

Once the eyes had dried a little and the crying stopped, Mike knelt next to the bed and clasped his hands together. Mary bowed and clasped her hands together also. Mike prayed, "Dear Lord, thank you for blessing the hand of Dr. Kissling. Thank you for sparing Mary's life and allowing the cancer to be removed. We humbly accept your will, Lord, and place our lives in your caring hands. We pour out our grief to you, oh Lord, for it is too much for us to bear. We know you have a purpose for all things. Make us worthy to receive your blessings. Amen!"

The traveler waved his hand over the couple after hearing their plea for comfort, and Mike and Mary looked at each other and smiled as they could feel some of the grief being lifted and hugged again. John the traveler smiled at them and then muttered, "And so it begins!" and he faded away with another cool breeze filling the room. Mike and Mary ignored the cool breeze, thinking that it might be something to do with the air-conditioning.

Other members of the family took their turns visiting with Mary, showering her with hugs and kisses and condolences until it was time for her to get some rest and the nurse pulled the door shut. Mike stayed in the room with Mary and fell asleep in the chair next to her bed.

CHAPTER 4

The Vision/The Birth

Mary recovered very quickly and was released from the hospital two days after surgery with limitations. She spent several weeks at home recovering, her strength growing every day and letting the stitches heal. The emotional scars would take longer. She went back to work at the bank to take her mind off what she had lost. She thought it would be better to be around co-workers and friends.

Two months to the day after the operation, Mary and Mike were aroused from a deep sleep by an angelic voice singing. They were startled and confused as they wiped their eyes and sat up in bed together. At the foot of their bed was the bright image of a beautiful woman. She was singing an enchanting song they had never heard before. They seemed unafraid of the event and remarkably calm.

The woman's appearance illuminated the entire room with a brilliant light. It was warm and comforting. She wore a white, flowing robe tied at the waist with a long, golden cord. Her golden hair waved back and forth as though she were facing the wind, but there was no wind. Her face was very pleasing to look at. She wore a crown of stars over her head.

She extended her arms toward them slightly with her palms up. Immediately, the room filled with the scent of beautiful roses, and she spoke to them. "Fear not! I bring you glad tidings from the Lord. He has felt your pain and heard your prayers. Mary, you will bear a girl child on the twenty-fourth day of December 1995. You shall call her Sarah! Alleluia! Merciful is the Lord God Almighty for he hears the cries of his children. Amen!" The beautiful angel shrunk slowly into a solid ball of light and shot upward through the ceiling. The scent of roses remained in the room, and the song continued as a quite lullaby, soothingly putting them back to sleep again.

They woke the following morning, singing and dancing, hugging, and kissing, happy with life, fully aware of what had occurred during the night. After breakfast, Mike rushed off to work, wanting to share the good news with his friends there. Now some people had a little difficulty talking to those who saw angels, and Mike found out the hard way that percentage was high. His boss accused him of drinking before coming to work. His co-workers and friends thought this was a mental thing that happened when people lost the ability to have babies. Either way, it was not the support he was hoping for.

By the end of his workday, Mike was convinced that he had not seen anything at all and that he needed to set an appointment to see a psychiatrist. That is, until he got home and found Mary still dancing in the kitchen. He realized that this kind of revelation was best shared with her. Over dinner, they got a real chuckle over the responses of their friends and some family members.

Mary and Mike were so happy they could not express it in words. The next several years were spent in anxious anticipation of the special event. They thanked God every day and attended church with a revitalized flair for worship. They spent quiet time in prayer and meditation. Mike spent many hours remodeling the second bedroom into a nursery with pink highlights. Mary spent much of her spare time knitting and sewing clothes, blankets, and cuddly animal toys.

Then one afternoon, during a routine medical exam in early May 1992, Dr. Kissling was shocked at the results. He sat down on the rolling stool and called for his nurse. The nurse peeked into the exam room, and Dr. Kissling directed her to have Dr. Plum and Dr. Andrews to meet him in his office as soon as possible. He excused himself from the exam room. When they arrived, they were slightly winded from trotting over to his office. Dr. Kissling asked them to look at the results from his exam, feeling that he made a mistake and they would be able to point it out to him. Dr. Plum and Dr. Andrews reviewed the results of the test repeatedly, trying to figure out where the mistake was made. Finally, the three doctors agreed that the exam results were conclusive.

All three doctors entered the exam room where Mary was looking at them, surprised and puzzled. Dr. Kissling introduced the other doctors to Mary and explained that what he was about to tell her would seem unbelievable. Mary nodded, patiently waiting for him to provide the news she was already expecting confidently. "Mary! I do not know how, but all my tests have confirmed that you are pregnant. I know this sounds impossible, but I had my colleagues here confirm the results." Dr. Plum and Dr. Andrews stood speechless.

Mary smiled and said, "I know. Isn't it great?"

Dr. Kissling half-smiled back at Mary and asked, "How is this possible?" hoping she would share some secret herbal formula with him.

Mary sat up and explained, "God did it. He promised me a baby—a baby girl to be exact—and here she is, right on time."

Dr. Kissling sat down at his desk in the exam room and wrote out several prescriptions for pre-natal vitamins and a schedule of follow-up visits. He handed the scripts to Mary, still mind-boggled as to what he was witnessing. She thanked him for the good news and bounced out of the office yelling, "Thank you, God!"

Mary left the doctor's office and went home to prepare a special dinner for Mike before she broke the news to him. She knew he would be excited and wanted to make it a special event for both.

Dr. Kissling spent the next several months consulting colleagues about Mary's case. None were any more informed than he was. They could not understand how a woman could give birth to a child after having a hysterectomy, reaffirming to Dr. Kissling that it was medically impossible. Still, Dr. Kissling's patient was pregnant and going to have a baby sometime in December.

On December 24, 1992, at exactly 11:59:59 p.m., an incredibly special package arrived at Desert Samaritan Hospital in Mesa. Sarah Ann Victoria Evans was brought into the world against all medical odds. She was as healthy as any newborn baby could be and weighed seven pounds and fourteen ounces. The nurses took baby Sarah and cleaned her up a little and placed her into Mary's arms as

Mike watched. Mary admired her new daughter as any proud mother would and motioned for Mike to come over. Mike leaned over and kissed Mary and Sarah as a tear rolled down his cheek.

Unnoticed in the corner of the delivery room was the figure of a man dressed in a tan robe and sandals. It was John the traveler smiling at the couple and their newborn baby. This time, there were two other witnesses with him, Tao and Myung-Dae, smiling and looking at the couple then at John. Floating above Mary was an angel, offering a blessing to the couple and Sarah. "As foretold in the sacred scriptures, a child will be born through a barren vessel. Witnessed by three, this child will be called Sarah and will bring forth a special miracle for the world of men. The Lord's blessing is upon you, so says the Lord God Almighty."

The scent of roses filled the delivery room, and the doctor and nurses commented back and forth. "Do you smell that?" one nurse asked.

The second nurse acknowledged her with a nod, and the doctor remarked as well, "Where did those roses come from? They're not supposed to be in the delivery room."

One of the nurses responded to the doctor, "There are no roses in here, but that smell is so heavenly." At the blink of an eye, the angel and witnesses vanished, leaving the room filled with the smell of roses.

Hordes of reporters waited outside the hospital, hoping to be the first to report on what they were referring to as the miracle baby. Four months into Mary's pregnancy, reporters, and medical journalists had picked up on the story and had been following up on it since. Special broadcasts interrupted normally scheduled television programing across the United States and around the world. Mike and Mary shot to celebrity status with Mary's last "Push!" and their little Sarah suddenly took her first breath and touched the hearts of nation and the world.

Mike and Mary knew that smell and felt comforted by it, knowing the miracle birth of little Sarah would bring lots of attention from people all over. They knew that she symbolized hope for people all over the world that miracles were real. This would strengthen people's faith in God's love for the world. It would turn some disbelievers into believers, and they were happy to share their joy.

Finally, after seven long years of anxious anticipation, they had a family of their own. They were so brokenhearted after Mary's bout with ovarian cancer they thought happiness was just an illusion. Then their hopes were rekindled by the visit of an angel. And now this! They went from face down in the gutter to king and queen of their own castle, excited to see their bundle of joy prancing up and down the hallways and playing in the room Mike had finished for her. It was a phenomenal triumph of true Christian faith.

Mike and Mary were so happy. Mike danced with the nurses and doctor, while Mary waved her arms and clapped her hands in the air to the imaginary music. After a quick celebration, Mary dozed off; she was still feeling the groggy effects of the epidural administered by the anesthesiologist. But Mike could not stop joyously celebrating and continued his party on his way to the waiting room, where the family was waiting.

After sharing the good news with the relatives, he excused himself and went to be with Mary. He sat down beside her and watched her breathe through two thin plastic tubes in her nose. He grabbed a tissue from the box on the nightstand and carefully dabbed the drool rolling down her chin from the corner of her mouth. Mike thought to himself, *It's times like these one curses oneself for failing to bring the camera.*

Mike was a very gentle-natured person. He was thirty years old and made a decent living in construction. Years of being in the Arizona sun left him with a dark tan on the exposed portions of his body. He had developed some serious, well-defined guns and strong legs. His face was lean and chiseled with high cheekbones and a square chin. He now had short, neatly trimmed, brown hair that had been lighted by the sun and coffee-brown eyes. As Mary woke, he smiled (he still had all his own teeth).

"You did it, sweetheart," Mike said softly while patting her head. "You gave us a beautiful baby girl. I am so immensely proud of you right now I could burst."

Mary was still a very stunning woman. She was twenty-eight years old with long, brown hair and brown eyes. She stood about

5'9" tall with a petite, firm frame, measuring 36-22-34 before the baby. She too had a beautiful tan. Mike sat there admiring her; she showed little evidence that she had just given birth to a seven-pound, fourteen-ounce child.

"We did it, honey. You and I make a great team together," Mary replied smiling.

"Merry Christmas!" Mike said as he handed her a little wrapped box.

"Oh, Mike, all your presents are at home." She sighed, taking the gift from him.

"Don't be silly, you have given me the best present I could have ever hoped for," Mike said and paused for a moment. "This is just a little something special to remember the moment by. Come on! Now open it!"

Mary grinned and quickly opened the gift. Inside the tiny box she found a golden charm bracelet with a little girl face charm on it. It read "Sarah Ann Victoria Evans, Born 12-24-92." Mary teared up and pulled Mike toward her, giving him a hug and a kiss. "Mike! It is beautiful… I just love it!" Then she urged him to place it on her wrist.

"Merry Christmas, darling," Mary said to Mike while looking around the room. "Where is Sarah? I want to see my baby."

Mike softly replied, "The nurse is cleaning her up and said she would bring her in shortly." Mike leaned in and kissed Mary on the forehead. "Would you like some ice?" He reached over to a blue pitcher on the end table and scooped some crushed ice into a cup.

Mary responded, "Yes, thank you." She smiled while bobbing her head up and down like a child would do for ice cream. She took the cup of ice from him and poured some into her mouth. She closed her eyes and mouth and let the ice sit on the back of her tongue and let the cool water run down the back of her throat. "Ice never tasted so good!" Mary replied with her mouth full.

The room door swung open, and in stepped a nurse with Sarah wrapped in a tiny blanket. "Guess who is hungry?" she said, walking over to the bed and gently handing Sarah over to her mother. "Here is your mommy and daddy," Angie said in a sweet little voice. "You may want to feed her now; she is pretty hungry. I will be back in a little while. If you need anything, just push the call button on your bed handle." Then Angie left the room, closing the door behind her.

Mom and Dad were beaming with joy. Sarah was quiet as she stretched out her arms over her head and yawned in her mother's arms. She squirmed back and forth for a moment and then made a face like she was going to cry, but nothing followed as though she had changed her mind at the last moment. She struck a pose and stared up at Mary. Mary checked as she watched Sarah dance around to find the most comfortable position.

"Oh, Mike, she is so precious. I cannot believe she is finally here. Look at her tiny hands and feet—they are so adorable. I love you, Sarah," Mike observed as Mary checked out her gorgeous bundle of joy.

"Once again, honey, your observation is flawless. She is Daddy's little girl. I love you, Sarah. Wait till you get home and see what I made for you," Mike said softly but swelling with pride, knowing the hours he spent preparing the perfect room for his baby girl.

Sarah wiggled, kicked, and stiffened both her legs. The she opened her mouth to let out a cry. Before she could utter a chirp, Mary snuggled Sarah next to her breast and let her feed. Mike and Mary watched proudly, witnessing one of life's greatest moments as Sarah suckled life-giving nectar from her mother.

Sarah fed with fervor, locked to Mary's nipple. "Does that hurt?" Mike asked.

"No, not really, just a sense of firm pressure," she answered, looking up at him. After several minutes of fervent feeding, Sarah slowly drifted off to sleep in her mother's arms. Mike and Mary knew that these were the moments in life you never forget. They made themselves comfortable, held hands, and watched in silence as Sarah slept.

Eventually, Mike leaned back into the chair next to the bed and fell asleep. Mary lifted her eyes upward and gave a quiet thanks to God for this miracle. A single tear rolled down her cheek. She knew how inconceivable this was. To give birth after a hysterectomy was impossible without the help of a miracle. She also knew that Sarah was part of a bigger picture, but she had no clue what that might be. Mary was excited and scared of what was to come next.

Just then, the nurse peeked her head in the room to see how things were going, and Mary just smiled at her. She smiled back and closed the door. Forty-five minutes of peace and quiet passed before Angie returned to the room. She walked over to the bed where Mom and baby were and said, "Dr. Kissling wants you to get some rest. I will take Sarah now, and when it is time to feed her, I will bring her back."

Mary nodded affirmably and surrendered Sarah to Angie, having trouble keeping her eyes open. Mike was snoring in his chair. The rustling in the room did not even cause a twitch out of him. He was dead to the world. Mary fluffed her pillow and sank down into the bed and fell asleep instantly. Angie took Sarah out of the room slowly and carefully so as not to disturb either.

After four hours passed, they repeated the ritual that would continue for several weeks. Mike awoke and walked over to the bed and kissed Mary on the cheek. "You okay, honey?" he asked, wiping the sleepers from his eyes.

"I am doing just fine, darling."

Mike continued, "And how is our precious little Sarah doing?" as he gently stroked her arm.

"She has a very healthy appetite and starting to get some tone to her skin color," Mary said, referring to the pinkish hue of Sarah's skin.

"You are right. She looks a lot better now. Before, she just looked like a raisin, but now she looks like a kidney bean," Mike said.

Mary fake-slapped him and said, "Do not refer to our baby girl as fruit and beans!"

Mike smiled.

Mary asked, "Are you hungry? Because I am starving."

He looked at the empty tray. "Come to think of it, I could go for a thick western omelet with extra cheese and a tall glass of orange juice," Mike said mockingly.

Mary brushed him off with a wave of her hand, thinking he was recovering from too much sleep. "Why don't you go down to the cafeteria and get something to eat?"

He turned and smiled at her while walking to the door. "That is a good idea. I will check with the nurse on my way and see about bringing you something too."

"Aw, who's my hero?" she said as she eyed the end table, where the container of ice had turned to water.

Mike flagged a nurse as he walked past the nurses' station. "Could you see that Mary gets something to eat? She's pretty hungry."

Angie nodded. "I will take care of it, Mr. Evans." Then he continued to the elevator and went down to the cafeteria.

Shortly after Mike had requested food for his wife, the door to her room popped open and an orderly brought in a tray of oatmeal, toast, coffee, hardboiled egg, and a strawberry yogurt.

Sarah had finished eating and went to sleep again in Mary's arms. She watched her sweet baby, admiring how quietly and peacefully she slept. She could not get over how beautiful she was and wondered what or whom Sarah would be when she grew up.

She pulled the breakfast tray over to her and positioned it so she could nibble on her food.

Mike ran into Dr. Kissling while exiting the elevator, heading toward the cafeteria. "How is Mary and Sarah doing, Doctor?" and before he could reply, he asked, "When will they be able to go home?"

Dr. Kissling looked up from the charts in his hand. "Hello, Mike! Well…if everything checks out okay, as I believe it will, you can take them home tomorrow." He started to walk away and then paused, turning back toward Mike. "Mike! I cannot tell you how privileged I am to have seen this for myself. I never would have believed this possible if I had not seen it for myself."

"Well, Doctor, join the club. I know that Sarah is a miracle from God. But sometimes I find myself asking why us?" Mike confided in the doctor.

"There are a lot of reporters outside waiting for you and your family. Remember, the only thing that is important right now is you and your family."

Mike walked up to the doctor and awkwardly shook his hand while attempting a bro hug, collapsing the charts against the doctor's chest. "Thank you, Dr. Kissling, for everything. Mary and I are very blessed to have you with us."

Dr. Kissling shook his head. "I don't know if I share your confidence, but nonetheless, I'm very pleased with the results." He turned and headed down the hall, yelling back at Mike, "I'll be up in an hour or so to check on you two… I mean you three." Then he continued down the hall, glancing back at his charts.

Mike shoveled a western omelet into his gut with a tall glass of cool milk. Then he rushed back up to be with Mary when the doctor arrived.

When Mike entered Mary's room, he found her sleeping, and the baby was gone. He made himself comfortable on the chair next to her bed, a chair Mike was becoming too familiar with.

After about ten minutes, the door opened, and Dr. Kissling entered the room. He whispered to Mike, "Hello, Mike. I came up to see how my favorite patient is doing." He walked over to Mary and started taking her vital signs. Mary woke up with a few moans and groans.

"How are you feeling?" he asked in a pleasant tone. "I mean other than that feeling of having passed a bowling ball through your legs." Dr. Kissling smiled sympathetically.

"I'm sore, still hungry, and I have a flabby tummy now," she replied while gently shaking her tummy fat at the doctor as though implying, *What are you going to do about this?*

Dr. Kissling scanned her chart and glanced over the top of it. "So everything is fine then!"

Dr. Kissling put the chart down, walked to the side of the bed, reached out, and held Mary's hand. "I'll have the nurse bring you something special, and we'll get those tubes out of you. You do not need them anymore. How does that sound to you?"

Mary smiled. "That sounds great!"

"Good! We agree. How are you holding up, Mike?"

Mike turned toward the doctor. "I'm fine. I just want to hurry and take my babies home and get them settled in."

Dr. Kissling nodded with an understanding smile. "We can do that tomorrow morning, okay?" Mike and Mary acknowledged by nodding together.

"There is one more thing we have to talk about. I have examined Sarah thoroughly, and she is one of the healthiest babies I have ever delivered. However, for some unexplainable reason, she is 100 percent deaf in both ears and mute. She cannot hear a thing, nor will she be able to speak. I am really perplexed. This is the first case of its kind…go figure! She has all the equipment, and it is all in the right places, but she failed to respond to all audio and vocalization tests. I just don't know what to make of it." Dr. Kissling scratched his head.

Mike and Mary both took a deep breath and let it out slowly then another, trying to stay composed. Mike embraced Mary and asked Dr. Kissling, "Is that all? Just that she will need to be taught how to communicate differently or is there something else?" They stared at the doctor with great anticipation.

"No, that is the extent of it. I am sorry. When you feel up to it, I will have some information sent up to you on how to cope with a deaf child. I will get you the name and number to an expert in the field who can help you." He waited for a few moments, gauging their response.

Then he added, "Because Sarah was born without being able to hear, she would not know the difference. To her, this is normal. With only a few restrictions, she will be able to lead a normal, healthy life." Again, Dr. Kissling waited to see what kind of responses he would need to deal with.

Mary looked at Mike then looked up at the ceiling and then at the doctor. Mike replied calmly after witnessing Mary's gestures, "Sarah is a blessing from our heavenly father. He has seen fit to guard her from those that would blaspheme his holy name. We believe that he has an incredibly special reason for this, and she is just as she should be. We are happy to have a healthy daughter. Thank you, Dr. Kissling, for being so kind and taking such good care of our baby."

Mike paused for a moment then continued, "Learning to speak without words will be a new experience for all of us, and we look forward to the doors it will open." Mary and Mike embraced each other again.

Dr. Kissling was absolutely humbled by their response. He had never been in the presence of this kind of faith before and did not know how to respond.

Usually, his patients blamed him, cried themselves sick, and sought legal counsel. Dr. Kissling was moved by the experience and walked to the bed where the two were embracing. He leaned down and mumbled, "If only I could inject all the patients I have with your faith and strength, what a beautiful world it would be. I am utterly inspired at this moment." He hugged them and wished them an extremely healthy and happy future together.

Dr. Kissling wiped a tear from his cheek and walked over to the door. As he opened it, he winked at Mary. "Now! It is time for that special treat I promised!" And he headed out of the room.

A few moments later, Angie entered the room with a big piece of chocolate cake. "Dr. Kissling said you might enjoy this, Mary." She walked over and placed it on the food tray already there. "Enjoy!" She walked out of the room.

The following morning found Mary fighting with Nurse Angie and Mike. She did not want to be wheeled out of the hospital in a wheelchair. Mike kept putting Mary back in the chair, trying to calm her down, while Angie explained sternly the hospital policy requires that she be wheeled out for her own safety. Finally, progress was made, and they started down the corridor as Dr. Kissling turned the corner.

"I am glad I caught you before you left. Mary, a few of my colleagues and I were wondering if you might come back in a few weeks to undergo a few tests. Your remarkable achievement recently has left us completely stumped. We are hoping you will help us try to understand what happened in your body that allowed you to give birth to Sarah. Please, Mary! It might help us to unlock a medical breakthrough that could help other sterile women in the future. It is completely voluntary and would only require you to stay one night with us… What do you say?" Dr. Kissling pleaded with her in a "Do me a favor," charming voice.

"Dr. Kissling, we are so thankful to you and your staff for every-thing you have done. Believe me, if I thought for one moment that your team of medical professionals would be able to find a shred of helpful information, I would gladly let you cut me open now… but medical science has never been able to prove or disprove a single miracle of God. I do not want this wonderful event clouded for one moment with a single doubting Thomas," Mary replied politely with Sarah bundled up in her arms.

"Tests will not be necessary, Dr. Kissling. We know that Sarah is here by the will and grace of God. There are no tests that can explain an act of God…so Mary will not be coming back for those kinds of tests," Mike relayed in a calm voice and with a warm smile.

"Very well then, I understand perfectly. I will explain this to my colleagues, and here is Dr. Collins's card. All the information you need to set an appointment with him is on the back of the card," Dr. Kissling said while handing the card to Mike.

"Drive carefully, and I will see you and Sarah after the first of the year for a follow-up. My nurse will call and set up the appointment. If you have any questions or concerns, do not hesitate to give me a call… Have a healthy and happy New Year! All of you!" Dr. Kissling shared as he turned and headed toward another nurse, who had been trying to get his attention.

Yelling back at Mike, he warned, "Be mindful of the reporters outside. 'No comment!' always works best for me."

Mike, Mary, and Sarah exited the hospital, surrounded by their parents and relatives, into a horde of reporters, cameras, and microphones, each scrambling for the first pictures and words from the miracle mom. Mary was now walking and thankful to be rid of that wheelchair. She slowly walked up to a podium set up for them next to the entrance of the hospital with Mike beside her.

Reporters and TV news crews had been waiting for this moment. An eerie hush fell over the crowd as Mary leaned forward toward the bouquet of microphones taped to the podium.

"Sarah, Mike, and I would like to thank those who sent the flowers and cards of well-wishes. It was truly kind of you to take us into your hearts like that. We found them very warm and comforting. Sarah is a wonderful blessing from God, and she is living proof to the world that God loves us all. He also wants us to be comforted in this wonderful miracle. He knows our hearts and answers our prayers. I want to express to those who are at a point of total despair that Sarah is your hope too. Her very presence here today is a true testimony of God's love for the world. Little Sarah stands as a bright beacon of hope, love, and glad tidings to all that live in the fog of chaos, hate, and despair. Thank you all for your friendship and support. May God bless you all."

Mary looks at Mike with his jaw dropping to his knees and said quietly away from the microphones, "I have no clue where that came from, but it sounded really nice." Mike agreed with a nod and hugged them both.

The reporters started with a barrage of questions as Mary turned and exited the podium. Mike stepped up to the microphones. "You have all been very patient, and we thank you. Mary and Sarah want to go home and get settled in now. And that is where we are going now, so please take a moment today and give thanks to God for all his blessing in your life and ask what he wants you to do in your life. Thank you."

Mike, Mary, and Sarah exited the podium area and headed to an awaiting car, being driven by Mary's dad. They entered, and the vehicle moved away from the scene as the reporters finalizd their thoughts and took their final pictures.

CHAPTER 5

Cairo, Egypt, June 1999

The sun was rising over Cairo, Egypt. It was a beautiful sunny day with a bright blue sky. Tim Wilford (an archeologist and curator for the Egyptian Historical Museum in New York) and Haseeb Jaffar (partner, friend, and owner of a local museum and curio shop) frantically rummaged through Haseeb's museum office. The museum was located near the trader's market and was filled with religious artifacts from various digs they had sponsored in Egypt, items of interest from all segments of history.

Tim and Haseeb were locked in a desperate search for a shipping manifest. "Probably the most significant document in religious history—and we go and lose the damn thing!" Tim huffed sarcastically as he sorted through a dusty box of papers at his feet, a perfect square box that held fifty light-brown hanging folders when full.

Unfortunately, Tim thought to himself, few of the folder tabs are filled out and those that are have the wrong label on them. Haseeb had badgered Tim about hiring a student from the college to help with the filing system for years. But Tim insisted they did not need help. Now Tim was giving serious thought to Haseeb's recommendation.

"We have not lost it!" Haseeb barked at Tim as he glanced over the oak rolltop desk covered with stacks of records. "We have simply misplaced it for the moment. Besides, if it is meant to be found, we will find it!" He appeared to be calm and collected with his emotions. Inside he was frantic and begged Allah to let them find the document swiftly.

"With the luck I have been having lately, I would not be at all surprised if someone was using it for firewood," Tim scoffed, frustrated at their failed attempts to find it. Tim pulled another box between his legs and dug into the files. A beam of sunlight danced across his eyes as he hunched over to sort through the file box. He slid his five-wheeled $89.95 market special leather-back chair over to avoid the beam. He could feel a migraine headache coming on and was becoming sensitive to the sunlight.

"Oh, stop whining, Tim! Beating yourself up is not going to help, so shut up and keep looking—we will find it!"

Last night, Tim saw something odd while he was examining the documents of a shipment sent to the New York Museum. He uncovered a cash of valuable scrolls during their last trip to a dig just northeast of Cairo. Tim and Haseeb had discovered a total of thirteen scrolls at the dig. All their research indicated that there would be only twelve scrolls there. He thought, *How peculiar that there are thirteen scrolls.*

Tim believed that the thirteenth scroll was written by the hand of John the Baptist. Written AD 1 just after Christ was crucified on the cross. He believed that the scroll would reveal the details of a miracle that would soon occur. What made it more fascinating was that John the Baptist was beheaded by a jealous queen before Christ died and was raised up, which means that John the Baptist would have been raised from the dead as well.

Another fascinating observation that Tim discovered while reviewing ancient writings from another dig was that John the Baptist may have been only one of many raised with Christ on the third day after his death. This opened a whole new realm of possibilities in religious discoveries and tied many missing pieces together, especially the writing discovered at the Dead Sea that many theologians dismissed as fabrication or writings insignificant to religious studies.

Tim was the opposite of most people. He derived his most brilliant ideas when he was sleep-deprived as he was now. Hearing the sounds of a waking city mesmerized him. He could hear the hustling and bustling of the early morning traffic in Cairo outside the museum. People walking by, an occasional horn blowing—it was all part of the beautiful symphony called the city of Cairo. Tim allowed his mind to wonder back in time.

For several years, Tim and Haseeb had been working the dig northeast of Cairo, Egypt. It was discovered when Tim answered the call of nature. They had a small dig going there after someone had discovered ancient human remains. The dig itself had nothing to reveal except the body of a man who had died in the desert—a nomad perhaps who had lost his way in the desert and died. Certainly, nothing to warrant the cost of additional resources.

Portable toilets were an unwarranted expense for this find and were not requisitioned. So one afternoon, Tim received the call and headed into the desert to answer it. He wandered away from the dig about one hundred yards to the north. His eyes danced across the desert landscape, watching the rising heat distort objects near and far. He stopped and relieved himself on the desert floor. He could hear a slight breeze as his portable poor boy hydro drill poured out in front of him.

His eye caught a glimpse of an odd shape protruding from the flat desert landscape beneath the end of his pee stream. At first sight, it looked like a pointed rock. But as he examined it more closely, allowing his urine to act as a small water cannon, he noticed it was more symmetrical. He stopped peeing in midstream and yelled to his crew. He was pissing on the very top of a pyramid, buried for thousands of years by the sands of the desert.

Tim thought, *How phenomenal the discovery and how crude the tools used. Haseeb had shared the story at every bar in Cairo and referred to his partner as Tim with the Divining Rod after the first drink. And rightfully so—his was the most important discovery they had found in their fifteen-year partnership. Haseeb had insisted that Tim drink lots of water on all their dig sites since this incident.*

It took them several weeks to remove enough sand from around the pyramid-type structure to discover it was relatively small in comparison to the Great Pyramids in Giza. It was only 255' high and 375' at the base. The sides of this marvel were exquisitely smooth unlike any other pyramid discovered so far, and at the base, they discovered human bones.

Tim remembered examining the groups or piles of bodies and noticed they all had very strange wound marks on their bones. Some of the bodies were in pieces. Others had large claw marks on their bones like that of a large bear. Tim and Haseeb became genuinely concerned over the safety of their crew and placed armed guards around the dig.

Once the earth had been removed, they discovered there was no discernable entry into this small pyramid-type structure. The hieroglyphics were unreadable. They looked like they had been scratched off or chiseled away. At the base of it, where the bodies were discovered, there was faint writing on the walls. But the dye or ink of some kind was scratched away and left the messages illegible.

Haseeb and Tim spent several hours trying to discern one of the messages that still had a few of its letters intact. "Do--- --en -is -llcer---- ---th" was all that they had to work with. Tim noticed there were six holes drilled or cut into the north wall shaped in an unusual pattern. The holes seemed to hold the answer to entry into the tomb or temple, he thought. Haseeb studied the stained or painted characters more closely and discovered what they thought was ink or paint was human blood.

After studying the holes for six hours, Tim walked to the wall, spread the finger on his right hand, and placed them knuckle deep into the first five holes. Then he placed his left index finger into the sixth hole and pulled the panel. To everyone's amazement, a large stone face next to Tim moved outward about six feet and stopped. The vacuum seal sent dust into the air as the stone moved and the air hissed around the opening. Like pressure being released from a truck's air brakes.

Then the hissing reversed and pulled sand into the opening for a moment and hissed again as though it were breathing like a huge animal. The foul odor forced Tim and Haseeb back from the opening about fifty feet. They took deep breaths of fresh air and forced them out. "Oh, my Lord, what is that hideous smell?" Tim asked Haseeb. He could only shake his head and shrug his shoulders as he fought for more air, trying to clear his head.

"It smells as though something is still in there and forgot to die," Tim said as he took several deeper breaths. A small boy stopped dead in his tracks as he approached the small pyramid-type building with her father. He dropped his things in the sand and stood motionless, staring spellbound at and around the structure.

The little boy was Amid, son of Habib, one of the supervisors on the dig. He saw a great, ugly beast trying to crawl out from the small opening, struggling to fit through the small opening as though it was too small for his enormous size. The hideous creature reached for one of the straggling workers heading toward the others that had already backed away. He felt the passing of the monster's claws against his back and turned to see what was there.

As he turned to face toward the pyramid the other workers saw that his garments were shredded on his back. They yelled at him to hurry and join them and not to turn back to the tomb. "Hurry! Masuda, hurry! Come this way and join us!"

The workers with Tim and Haseeb saw nothing near the structure entrance, but they recognized some sense of danger was over there. Masuda turned back toward the other workers and ran to join them. A second later, a plum of dust rose from where he was just standing. When he reached Tim and Haseeb, he turned back to look toward the pyramid with the others. They saw nothing but the remaining dust of the plum dissipate across the desert floor. But they heard heavy breathing and occasional hellish screams from the opening.

The workers who earlier had gathered around the pyramid as the event unfolded, applauded, and cheered as Tim successfully moved the stone forward were now debating whether to stay or leave. Amid grabbed his father's arm and pleaded with him not to approach. Habib asked his son, "What is the matter with you, son? It is my job. I must go and help."

Amid clung to his father, begging him not to go. "Do not go, Father, you may be harmed. The beast is angry and will be loose soon."

"Please, Amid, let me go! I must, they are waiting." The boy reluctantly released his father and fell into the sand crying. After a few moments, the boy knelt on his knees and prayed to Allah for protection and to spare his father from the beast. Habib approached the fellow workers and they surrounded him, asking what was wrong with his son. "I do not know, he may be touched by the sun. He was babbling about a huge beast he saw crawling out from the tomb."

All together the workers looked across the empty desert and around the tomb and saw nothing. They looked back to see Amid still on his knees, pleading to Allah to spare his father and the workers. They started moving farther away from the tomb slowly. "What is wrong with you?" Habib yelled at them, not believing the fear that showed on their faces. "All of you get back to work. The boy has a touch of the sun, that is all." Hesitantly, the men started back toward the pyramid.

Tim and Haseeb joined the workers; they were also concerned about the boy. Tim looked at Habib and asked, "What is the matter with your son?"

Habib responded, "Nothing! He said he did not want us to approach the tomb. He said there is a huge beast crawling out from the entrance."

Tim looked at Haseeb, puzzled. "Go see what you can make of it, please!" Haseeb walked over to the child and put his arm around his shoulder. "What is it that you see, son? Please be specific."

Amid stopped praying and inquired of Haseeb, "Can you see the beast that is crawling out of the tomb?"

"No, son! I do not see anything over there. But I want to know what you see there. Please tell me." The boy looked back, and his eyes opened wider; he started to shake with fear. Amid started talking in and ancient Masoretic dialect of Hebrew. He prayed for deliverance from the beast…pleading for the safety of the workers to the god of Abraham. Then he paused a moment and began to describe in detail what was unfolding before him. Haseeb listened to Amid very intently and somewhat startled.

"I see a large, ominous beast trying to escape from the tomb. He first appeared when Mr. Wilford opened the sealed door to the pyramid. The sound that is echoing from it is the monster's breath. It is a bad smell…of rotten flesh and screams of lost, burning souls.

The beast has the head of a jackal with rams' horns on either side. He has a great tail with a point on the end of it. He is very dark in color and looks like a skeleton with exceptionally long claws and has three clawed toes with great talons on the back of each leg. His eyes are like bright red, burning coals."

Amid breathed deep and looked back at Haseeb. With terror in his eyes, the boy quietly (almost a whisper) told him, "The monster is free and looking at us now." Haseeb motioned to Tim and the workers to move farther back. They complied and reluctantly wanted desperately to check out the booty in the tomb. Haseeb and the others noticed more dust being kicked up around the opening of the pyramid.

The boy gasped and looked to the clear, blue sky with a tearful smile. "Look, Haseeb! They are coming to help us." He pointed to the sky. Haseeb looked to where the boy is pointing and saw nothing. "Describe what you see now, Amid." The boy continued to speak in the old Masoretic dialect of Hebrew, a language the child has never been exposed to. "The beast saw them too…two angels dressed in white, flowing togas with golden breastplates and battle kilts. They are carrying bright silver swords that shine in the sunlight and are riding two beautiful white stallions with wings."

Tim, Habib, and the workers gather behind Haseeb and the boy as they listen to Haseeb translate the strange language the boy uses to describe the event as it unfolds. A gust of wind kicks up from nowhere, and four hoofed plumes of dust appeared between them and the tomb. The group gasped with uncertainty at the same time, unable to understand what they were seeing in the sand in front of them. The angel steps off his steed and kicks up two more plumes of dust.

He charged at the beast without warning as the beast reared up to strike. The angel strikes a blow to the beast's leg, and it falls backward over the pyramid producing a large cloud of dust that rises over the back of the pyramid. The servant of Satan pounced toward the Angel of Light and slammed him into his steed. The horse stands firm as the angel regains his balance. Then pushing forward, the angel lets loose a fury of strikes, so quickly it was a blur, but the boy could see the monster shrinking in size, falling to the ground in pieces, each piece being absorbed into the sand.

As quickly as it had begun, it was over, and the angel raised his sword in victory and praised God. Before hopping back onto his horse, he glances over at Amid. At once, the boy waved at the victorious Angel with Jubilance, laughing and smiling. The others joined in, not realizing what the boy saw but felt it must be good. The angel winked at the boy and jumped back onto his horse, the horse reared up as if to wave goodbye to the boy and rode into the wind, joining the second angel that stood guard to bear witness over the battle before disappearing into the heavens.

As the boy stood and waved, all the men in the group waived to the empty desert sky as well. They were not sure what had taken place, but they did witness the dust of the desert being stirred while Amid explained to Haseeb what was happening. The workers then bent down on their knees and bowed, praising Allah for their safe delivery (based only on what they were told by Haseeb). Tim just scratched his head, and Haseeb patted the boy on the head and thanked him for his vision.

Habib ran over to his son and hugged him tight, kissing him on the forehead then raised him to his shoulder as a hero, as the boy laughed playfully. After praising Allah, the men joined Habib and his son, dancing him around the pyramid calling him a man of vision. Habib took his son off his shoulders and placed a lamp in his hand, inviting him to join them as they prepared to enter the tomb. Tim glances over at Haseeb and Habib and nods approvingly. They start back toward the tomb entrance, eager to see what was so important that the angels in heaven and the demons of hell had to fight over it.

They lit their lamps and entered the small entrance to the pyramid slowly, carefully walking against the wall so not to disturb any of the dust that lay on the floor. Tim thought that would be wise because if there had been any foot traffic in the tomb after it was sealed, they would be able to see it when they brought in the big light. They traveled down a small corridor and stopped. Amid was in the lead, and the boy saw something that no one else could see;

he stuck out a stick and raised it up and down. Suddenly, there was a slight clicking sound, and several darts shot from the wall.

Tim tossed his hands up into the air and looked back at Haseeb and Habib, indicating he had no clue how Amid had snuck up to the front of him. He let out a gasp once the trap was sprung and patted Amid on his head. The boy turned his head toward Tim and smiled. They continued, entering a large room where they discovered two sarcophaguses. Elaborately decorated and depicting royalty. Tim called out to Haseeb, "Who do you think this belongs to? Must have been someone important." Haseeb shook his head, indicating he had no idea.

Haseeb tried to examine and read the hieroglyphics on the walls inside the main room, but he was unsuccessful. They had been tampered with and were not legible. "These writings have been chiseled away, Tim, I cannot read them." Tim glanced over to the wall that Haseeb was examining and noticed the chisel marks on the wall. "Well! Someone did not want us to know who was buried here for some reason." They continued, gently searching the room that was cluttered with gold, jewels, and artifacts.

Little Amid disappeared behind a stone-carved altar at the back of the room for a moment and then reappeared holding an old cylindrical case. "Is this anything?" He held it up for all to see. Tim and Haseeb rushed over to where the boy was standing and discovered a shelf behind the altar, displaying several scrolls. Tim took the leather-type cylinder from Amid very carefully and examined it. "Haseeb! Look at this!"

Then he handed the cylinder to Haseeb and turned back to the shelf to examine the other cylinders lying on the stone shelf behind the alter. Haseeb examined the cylinder case carefully and announced, "Tim! I believe we are looking at never seen before, writings of the Twelve Apostles of Jesus Christ," running his fingers across the leather container. "I will need to get these over to the museum to check them out in more detail, but I do believe this is incredibly significant. Look at how well-preserved they are!"

Haseeb lay down the first scroll on one of the sarcophaguses and turned to grab another from Tim as he handed it out from behind the altar. "Are you sure?" Tim asked with great excitement in his voice. "I can't tell for sure yet, but yes, I think we have just discovered the twelve scrolls!" Haseeb remarked as he leaned over to grab another cylindrical case from Tim.

After retrieving all the cases from behind the altar and lying them on top of one of the sarcophaguses, Haseeb counted them out loud, "one…two…thirteen."

Tim quickly stated, "Thirteen? I thought there were only supposed to be twelve. What is with the thirteen here?" They look at each other with eyes wide open and jaws dropping. Tim has a little bit of drool dripping from the corner of his mouth.

As they pondered over the scrolls, flashes were going off everywhere to capture each move made and to record photo evidence of where everything was discovered. There came a racket at the entrance of the tomb as the big light was carried in by two of the workers. They set the large lamp down carefully in a corner of the room on its own tripod and flipped the switch on the side.

The room was lit up brilliantly by the million-watt beam. Tim immediately noticed that twelve of the cases were made of lighter-colored leather skin, perhaps lamb skin, and one was distinctively different, perhaps goat skin because it was darker in color and with more pores. "Haseeb, look at this! This one does not belong here—it may be fake. It is much darker than the others and the skin is cruder."

Tim took the darker case and set it aside from the others. "These here are the original twelve cases," pointing at the twelve identical leather cylinders on the sarcophagus. "Yes, Tim, you are right. It is too easy. What was this guy thinking?"

They gathered up the scrolls and carefully encased them in a wooden crate to prevent any damage to them. They put two guards on the tent they laid the scrolls in, including the thirteenth scroll they were guessing to be a fraud. They finished taking pictures of the inside of the tomb to keep a record of all that was in it and where it was originally found when the tomb was opened. This was standard procedure for archeologists to keep a visual record for historians to study.

Back at camp that evening, after they had spent an exhausting day recording the artifacts, they carefully removed them to be crated and taken back to the museum. Tim and Haseeb sat in the tent where the scrolls were secured. They tenderly removed the thirteenth scroll from its container and Haseeb read through it.

Haseeb noticed that the dialect of the writing was the old Hebrew dialect spoken and written among the common people of Jordan, back in the days of Christ teaching. His attitude toward the scroll began to change significantly as he read the first portion of it:

> *I John, am known to all as the Baptist, put together my full recollections of my brother Jesus the Christ who comes after me. I set to words what I am to do and what will come after his death. To include the Miracle of Miracles that will take place in the first year of two thousand after his death.*

Hearing the words as Haseeb read them, Tim jumped to his feet and screamed "Jackpot!" then danced with Haseeb around in a circle, singing, "If I were a rich man…!" and kicked his heels together in the air. Haseeb had never seen Tim so excited about any previous discovery. "Why does this discovery make you so happy? We cannot sell it."

Tim stopped dead in his tracks and asked Haseeb, "What do you mean? We cannot sell it. Of course we can sell it! We can sell anything!"

Haseeb paused for a moment then turned to Tim. "No! We cannot. This is the first hope for mankind in nearly two thousand years. This scroll, if authentic, represents…the most wonderful gift to people of our generation from God through the hands of his servant. How in the world could we possibly consider selling this?"

Tim looked at Haseeb with rage in his eyes and screamed, "We can sell it…to the highest bidder…for his private collection. They will pay all the gold in the world for ownership of this document. You are not screwing this up for me Haseeb, so get off your high horse. You will spend the money just like I will and be glad to have it." Haseeb sighed and left the tent.

Outside the tent, Haseeb took a few breaths and let them out to calm down. He was enormously frustrated with the ignorance and greed that Tim was displaying. He had never seen Tim so fixed on gold before. Slowly, he scanned the desert's beauty. It was always beautiful at night. The stars shined bright in the black sky above. There was a cool breeze flowing through the camp. Everything seemed calm and serene.

He heard some chanting in the distance…a man next to the fire was chanting in a strange language…getting louder and louder. Haseeb motioned for the guard to get Amid, the son of Habib, and brought him over. A moment later, the boy was standing next to Haseeb. He leaned down next to the boy and asked him, "That man by the fire, who is he? And what is he chanting?" as he pointed to the man by the fire. The boy stared at him for a moment, then he stepped back behind Haseeb, just as his father, Habib, joined them.

"The man is unholy. He is chanting to draw forth the demons of the night. They are circling him overhead, and he is commanding them to take over our thoughts." Haseeb motioned to the guards to shoot the man by the fire. The guards reacted immediately without hesitation, and several shots rang out into the desert air. The man by the fire fell forward directly into the campfire, and his body exploded on contact with the flames.

Haseeb patted the boy on the head. "You are truly gifted, son. Allah has blessed us with your presents tonight. Thank you." He walks back into the tent to find Tim getting up off the floor and acting as though he was drunk.

"You feel better now, Tim?"

"Yeah! I guess, what happened? I cannot remember anything," Tim replied as he stabilized himself and sat down on the cot. "Nothing that could not be fixed by a gunshot into the head of an evil chanter."

"Wow, no shit!" Tim shared while rubbing the back of his neck. Haseeb sat down next to Tim on the cot and explained how he knew that the thirteenth scroll was the real deal. "Tim! You remember the research article in Archeology Today? About the parchment found in Israel in the early 1870s. Translated… It spoke of John the Baptist leaving the masses after baptizing Jesus and going into solitude in the desert. There he wrote his record of events about his purpose to pave the way for the Son of Man and he predicted his death. He also predicted a miracle to take place in the final days."

Tim nodded, indicating he remembered the article. "I believe this is the actual parchment that was referred to in that article. And if you recall, you showed great interest in the article written by Kim Lee in the South Korean newspaper in 1951 where the reporter uncovered the story from a boatman who explained that he accompanied a young monk to a burnt temple in Suwon, South Korea. That his mission was to acquire a sacred scroll wrapped in hide and transport it safely to the Tibetan Temple south of the Tibetan Plateau in China. The boatman spoke of a great battle between the monk and a demon."

Tim was looking very intently at Haseeb as he continued, "Then in 1974, there was the magazine article of a disillusioned Tibetan monk who told reporters of a strange visitor, tanned skin, long, smooth hair, and wearing a brown-colored camel skin robe with sandals on his feet. He had arrived at the temple one summer day in June 1971 to see the Elder, Tao. He stayed a brief time and then left with that same scroll…this scroll, Tim!"

Haseeb held it up in front of him and continued, "In that same article was a very interesting interview with a local man who was near the Temple at the time, teaching his seven-year-old son how to track. They say…a tall, amber-skinned man, with smooth shoulder-length hair, dressed in a camel hair robe, and sporting well-worn leather sandals…carrying a case in his hands was standing alone in the forest. He then turned his head for a moment to see what had captured the stranger's attention at the Temple, and when he looked back the man had disappeared into thin air, without a trace."

"I tell you, Tim, this is the thirteen scroll. This is the parchment that good and evil have battled over for two thousand years." Haseeb set the scroll down on the box in front of Tim. Tim shook his head, still doubting Haseeb. "You have presented a very compelling argument. But you have not provided any timeline correlation to this document. It was found with the other twelve scrolls in an Egyptian pyramid…explain how that scroll got into the mix without being there all along."

Haseeb grabbed the pictures taken of the tomb when they entered. "Look for yourself, the dust on the top of the twelve scrolls is the same density and color throughout the shelf that they rested on. Which means they were placed there at the same time together. Probably some thousand years after the tomb was sealed… I do not know how yet, but I suppose based on the subject matter, we can assume it was a spirit or angel. There were no footprints of any kind in the tomb before we arrived."

Haseeb handed him another picture taken of the other scroll. "You see how much thinner the dust is on this case…it was placed there very recently. And I will bet you it was placed there by the same man that showed up and disappeared from the Tibetan Temple in 1971." Haseeb smiled as he finished his argument. It was brilliant, accurate, and most of all provable (he thought).

Tim jumped off the cot, still staring at the pictures taken earlier. "You are right! We can prove it. The timeline is in the layers of dust. Look here!" He points to a picture that reveals the dust on top of the cylinders as well. "The dust is thicker on the top of these twelve scrolls and almost nonexistent on the thirteenth scroll. This is irrefutable evidence of the timeline of their placements into the tomb…if you also look here at the top of the original relics placed in the tomb at the time of burial, there is much more dust."

Tim heard the echo of a familiar voice in his head and returned to reality. He shook off his trip down memory lane. Now they had lost the most important find in history. They had to find this scroll, and they had to find it now. "Did you check these boxes yet?" Tim asked Haseeb as he pointed to the two boxes in front of him.

"Yes, I did. All those boxes in that corner have been checked already," Haseeb added. "The boxes on the desk to your right have not been checked." Tim grabbed the two boxes off the desk and began searching them.

Imagine, Tim thought, a miracle foretold almost two thousand years ago, taking place now, in my lifetime. What a discovery that would be…what a fantastic way to help unite the churches of the world together in a common belief. Tim had always felt that man had corrupted all the churches of the world…to serve their own needs.

Tim also felt, Religious leaders built themselves up and placed themselves loftier than the common man. Deep inside Tim believed that God did not intend for man to complicate religion. God, he thought, would want it to be simple so everyone would embrace it every moment of every day. And in so doing every living creature, including man would be able to live in harmony.

He often wondered what type of miracle would be powerful enough, in today's ever-changing cultures, to propagate a common belief among the masses. Would it be world peace? A cure for AIDs or…maybe…the second coming of Jesus? "Wow!" Tim blurted out loud. Then he shook his head and went diligently back to work searching for the darn miracle. Haseeb eyed him curiously for a moment after hearing his exclamation.

Tim could only imagine for now. He knew they had to find the thirteenth scroll and release it before the miracle took place. If not, the discovery of the scroll would not be near as effective. Their discovery had to be released before the event occurred to lend viable credibility.

Frantically they reviewed all the shipping documents. Possibly a careless recorder had accidentally shipped it to the United States. The search was exhausting, and beads of sweat formed on their foreheads. Frustrated from their inability to locate what they were searching for, both threw their hands up in dismay and cursed to beat the band. They scanned the mess they had made in the office and looked at each other hopelessly.

"Where the hell did that scroll go?" Tim asked Haseeb as he slammed his fist on the table. "Damn it! Do you think one of the workers took it home for private sale?"

"No!" Haseeb responded defensively. "My workers are very loyal to us. They would never do such a thing. That is why I pay them so well. I trust them."

Both men needed a rest; they took a seat across from each other, wiping sweat from their brow. Then took several breaths in and out to calm down and get their second wind. They were determined to find that document. Pondering what to do next, Tim saw the corner of a document sticking out from underneath the filing cabinet. He slowly bent down and lifted the corner of the now empty cabinet and removed the document carefully, reviewing it.

Haseeb noticed Tim's curious demeanor and asked, "What do you have there?"

"I'm not quite sure…" Tim replied, as he continued to study the document. He noticed it was a shipping document for the items sent to the New York Museum. It showed that twelve scrolls had been shipped among other items. As he reviewed it more closely, he discovered that though the order showed twelve scrolls were to be shipped, the attached receipt from the courier confirmed that thirteen scrolls were signed for and picked up for delivery to the airport. The order had not been amended to reflect this.

"God as my witness, Haseeb! I think I've found it," Tim exclaimed enthusiastically. "It's on the way to New York, to the Museum of Natural History." Excitement was mounting in his voice.

Tim stared through Haseeb, pointing and shaking his finger at him while formulating a plan. Then he unveiled his plan to his friend and partner Haseeb. "Call the airport and get me on the next flight to New York. Then call Carl Curry in New York and explain what happened." Tim paused for a moment, allowing Haseeb time to complete his notes and start dialing the airlines.

"Tell Carl not to open the containers until I arrive. Stress to him the urgency of this matter, but do not alarm him and do not say anything about the scrolls over the phone. I want to tell him myself, in person. He is going to shit a brick house right out of his ass!" Haseeb looked at Tim with a very pained look on his face. "I would not want to be there when you tell him. That's going to hurt like hell." Tim grabbed his coat and briefcase from the chair and headed for the door.

"I will take care of everything…you just get your skinny butt over to the airport pronto. The ticket will be waiting at the information counter. I will call Carl and let him know you are on your way and to pick you up at the airport when you arrive. I will be very discreet about the matter." Haseeb responded, very controlled and methodical, grinning with a huge baby smile and nodding affirmatively.

Tim reached for the office doorknob and grabbed hold of it to yank it open then stopped. He looked back at Haseeb, already on the phone with the airlines. "Haseeb?" he called out loud to get his friend's attention.

"Yes, Tim!" Haseeb said, looking up to catch Tim's delightful smile.

"We are about to make history, my good friend," Tim stated with the confidence and enthusiasm of a man about to fulfill a life-long dream.

"I honestly believe you are right, Tim," Haseeb said as the reality of his statement took hold of him. "May Allah protect you, my friend." Haseeb bowed slightly with his hands pressed together on his chest under his chin.

"And you, Haseeb." Tim returned the gesture and opened the door quickly.

Haseeb returned to his task and booked a ticket for Tim on the next flight. He was able to get Tim on a flight leaving in two hours. The flight would put Tim in New York late tonight. He then called Tim's apartment and left the information on his answering machine so he would have it when he arrived.

Tim had a taxi wait for him while he ran into his apartment to pack a few things for his trip and check his voicemail. He knew that Haseeb would get him the next flight. Sure enough, the message left on his answering machine confirmed his flight was leaving for New York in one hour and forty minutes. He grabbed his tan canvas overnight bag and stuffed it with a change of clothes, toiletries, his address book, and his Indian medicine bage that he carries for luck.

Within ten minutes, he was back in the cab. He told the taxi driver to take him to the airport as fast as he knew how. The taxi sped off. Tim arrived at the airport terminal just before noon. He paid the taxi driver and looked at his Swiss Army watch and noted his flight was departing in forty minutes.

Tim ran to the TWA information/ticket counter to pick up his ticket and rudely cut in front of everyone else in line. "Sorry, folks, this is an archaeological emergency. Doctor coming through… Make way please… Doctor coming through." Tim walked straight to the counter with his passport exposed for all to see. Of course, he kept moving it up and down, forward, and back so no one in the line could get a good look at it.

"Do you have a ticket for Doctor Tim Wilford?" Tim asked while trying to catch his breath. As he waited for a response, he hunched over slightly, resting his right arm on the counter. His head slightly tilted down while taking deep breaths.

Jessica, the customer service ticket agent, quickly whisked her fingers over the computer keyboard and pulled up his reservation. Then she confirmed his name with picture identification. She noticed on the card that he was not a medical doctor and started to question him about using a medical emergency to get ahead of the line.

"Dear? Before you say one word…let me say this." He was still gasping for air. "Those people behind me don't know that I am not a medical doctor. Now! After you have given me my ticket…do you want to be the one to tell them that?" Tim looked at Jessica with a puppy dog frown.

She smiled, checking him out. Seeing how pathetically out of shape he was but also how amazingly good-looking, she announced loudly so all could hear, "Yes, we do, Doctor Wilford. You are booked on TWA Flight 117 departing from gate 10. Your flight will be board- ing in twenty minutes. Good luck with your procedure."

Jessica leaned closer to him smiling and quietly stated, "You owe me dinner at a restaurant of my choosing when you get back, tough guy." She released her hold on the ticket, pointing the way to gate 10.

Tim said, "You got it, missy. I am all yours."

Tim turned and started to briskly walk toward his gate. A med- ical doctor stepped out of line, motioning to the counter. "Doctor? I am Dr. Reynolds from Stanford Medical Institute. What emergency awaits you in New York?" Tim glances at the spunky doctor in the crowd and clears his throat. "It is a scroll extraction from the anal cavity with complications." Then he stepped off lively, waving his ticket and asking everyone to wish him luck.

Comments from the crowd included "Oh, that is serious!" and "Poor fellow, I hope he makes it!"

Then Dr. Reynolds turned back to the line and paused a minute before responding. Everyone in the line was patiently waiting for the doctor to endorse his comrade in medicine. He looked at everyone in the line, shaking his head and raising his hands in the air and yells back to them. "He is a freaking dirt doctor, you know…an archeologist!"

Tim hustled all the way to gate 10, stopping long enough to get through a security checkpoint. As he finally approached the gate, the stewardess was just closing the door to the boarding ramp. "Wait a minute! I am on that flight!" Tim yelled. The stewardess reopened the door, checked his boarding pass as he entered the walkway, and closed the door behind him.

As he ran down the ramp and approached the airplane door, a steward was waiting with his hand on the door. "I'm sorry," Tim said as he boarded the plane.

"I see we are cutting it close today, sir," the steward, said responding with a smile.

"It was a last-minute reservation," Tim replied in between breaths.

"No bother, sir, nice to have you on board. We have heard of these new fad diets with a cardio workout and how it helps with the digestion system. You have, no doubt, flown on one of our meal flights before, I take it?"

Tim could see the steward was making a joke, and to be polite, he responded, "Why, yes, I have, my good man. Would you please inform the pilot I am on board and not to dottle with the takeoff? I am in a bit of a hurry today. Oh! And please inform the pilot that I am awfully familiar with the potholes on runway 3. Let us try to miss them this time—my hemorrhoids, you know!"

The steward smiled and closed the door behind him. "It will be my pleasure, sir, to relay those instructions to the pilot."

"Gabby! Please show Dr. Wilford to his seat and make him comfortable." He turned Tim over to the first-class passenger stewardess.

"Certainly! William I would love to." Gabby escorted Tim to his seat and took hold of his bag. "Would you like me to put your bag in the overhead compartment for you, sir?"

"Yes, thank you very much," Tim replied as he takes the window seat.

The plane quickly taxied to the runway and turned onto the main takeoff point. It immediately revved up its engines and thundered down the runway then eased into the air very gently. Tim removed his coat and asked for a pillow. Then he drifted off to sleep. He did not like to fly all that much because of all the plane crashes he had read about. So he coped with his anxiety by sleeping.

CHAPTER 6

New York Museum Curator

Carl Curry was at his penthouse suite in New York on the couch with his beautiful secretary, Anita French. Both were in a passionate embrace, sucking face. Twenty-four-year-old Anita was working her way through college at the New York Museum. She had long, curly brown hair and big brown eyes, which were accented by her dark complexion. She had a soft hourglass figure with large, sumptuous breasts, and she stood 5'6" tall.

Carl Curry was an oversexed thirty-five-year-old, who had found his niche in life. He was 6' tall with broad, squared shoulders. His complexion was pale with short, blond receding hair and blue eyes. He had a large frame with medium build.

Being the curator of a very prominent museum kept his nights free for extracurricular activities, most of which involved Anita. He lived a casual, carefree lifestyle due mostly to a large family inheritance.

Carl and Anita had been seeing each other since he hired her two years ago. He created her position when he met her at the museum. She knew all about it and preferred not to say anything. They were attracted to each other from the moment they met and had been going at it ever since.

"Carl, are we going to play nurse tonight?" Anita asked with playful anticipation.

"Not tonight, Anita," Carl said as he retrieved a small red bag from under the couch. "I have a better idea." He smiled at her. "Here!" He handed her the small bag. "Slip into this!" he coaxed.

Her eyes widened with excitement as she investigated the bag. She pulled out a black, skimpy G-string and two red tassels. "Oh, you kinky devil, you." Then she gave Carl a peck on the cheek and walked over in front of the fireplace.

As she did, Carl grabbed the remote control and adjusted the track lights, making the fireplace look like a stage. Then he pushed another button, and the music changed to something risqué.

Anita gently removed her clothes. She first removed one white fishnet stocking and then the next, stripping to the rhythm of the music. Carl watches with anticipation as she performed like a professional stripper.

She teasingly removed every piece of her clothing. Then she danced naked, running her hands all over her body. Her hands moved like they were the hands of her lover, taunting him with everything she had to offer. Slowly, to the rhythm of the music, she put the G-string on. Then she placed the tassels over her nipples.

Anita continued her seductive dance routine. With tassels tossing and her hips swaying, she inched her way over to the couch where Carl was sitting. Carl adjusted himself as his pants became more and more restricting. She bent in front of Carl, only inches from him, and started to swing her large, firm breasts toward his face. Carl leaned forward to meet them.

She pushed him back on the couch and dropped to her knees. He tries to lean forward only to be shoved back again. Anita forced his legs apart and began to rub the inside seam of his pants. She slid her fingers up the side of his waist and all over his torso, firmly massaging his chest and stomach.

Carl was hot; his erection was the size of a large, plump banana. As it continued to pulsate, it started to force its way through the top seam of his briefs. Anita slowly loosened his belt then undid his pants and slid the zipper down slowly and carefully. She gently ran her fingers up and down his bulging shaft through his underwear.

Carl closed his eyes and tilted his head back, surrendering to her completely. He moaned exotically as she pulled his pants down to his ankles and ran her fingers back and forth on his bare inner thighs.

Carl moaned louder, unable to control his excitement. He yelled, "Oh, yes! Yes! Anita, yes!"

Feeling properly coaxed, Anita placed both her hands on Carl's waistband and started to tug down his briefs.

Suddenly, the phone rang. On the third ring, the answering machine kicked in, and the message started to play: "*You have reached the residence of Carl Curry. Your call is important to me. After the tone, please leave your name, phone number, and a brief message, and I will return your call as soon as possible. Thank you.*"

"Carl? This is Haseeb Jaffar. Pick up! Tim Wilford told me to call you. It is especially important." Haseeb paused for a few seconds to see if Carl was going to pick up.

Carl picked up the phone and responded sharply, "This had better be really important, Haseeb! I was extremely busy doing some important research of my own." It was clear that Carl was frustrated.

"I have been trying to get through for several hours. The phone system is messed up. Anyway, have you received that shipment we sent to you?" Haseeb barked back in an anxious tone.

"Yes, I received that shipment today. We were going to open it in the morning," Carl replied, concerned at Haseeb's excitement.

"Do not open the shipment! I say again…do not open the shipment in the morning! Tim is on a flight to New York right now and should be landing"—there was a pause while Haseeb calculated the time difference—"in forty-five minutes on TWA Flight 117 from Cairo. He wants you to pick him up at the JFK Airport. He will fill you in."

"All right, Haseeb. I got it! I will go to the airport immediately and meet Tim. I will talk to you later, goodbye!" As Carl hung up the phone, he turns anguished toward Anita.

"Sorry! Honey! I need to pick up Tim at the airport. His plane is landing in forty-five minutes. We will get together tomorrow night, okay?"

"Sure, I understand. These things happen, but tomorrow you dance for me, tough guy!" Anita says as she walked over to Carl, slowly licking her lips, and playfully patted Carl on his firm, tight ass.

Carl moved toward Anita, hoping she would help release the throbbing in his pants. Anita knew exactly what he wanted, but she also liked to tease him. She looked into Carl's eyes with one eyebrow raised and whispered, "You had better get the lead out. It is going to take you forty-five minutes to get to the airport, and you do not want to keep Tim waiting…do you?"

"Shit! No! Damn it! You are right—I am going to have to bust some lights to get there on time," Carl continued to mumble aloud with his hands on his hips and his pants still around his ankles.

"By the urgency in Haseeb's voice and the fact that Tim is almost in New York, this has to be something particularly important…this could be huge!"

Carl sadly glanced down, pulling the waistband of his briefs out. "Sorry, little buddy, we have to go to work now. But if you are good in the car and you do not act up, Anita said she will come back tomorrow (whimpering) and we can play (more whimpering)."

Anita turns back to look at her pathetic boyfriend. She moved toward him this time, licking her full ruby-red lips. "Can I talk to him, Carl? It will just take a few minutes and…well"—she looked at Carl's enormous bulge—"I really do not think he is listening to you."

Carl looked at Anita thankfully and said, "You know him… sometimes better than I do…and I know he is going to act up in the car if you do not talk to him now. Besides, he always listens to you."

Anita dropped down to her knees, pulling Carl's briefs to his ankles, and…

(*Webster's 21st Central Dictionary* describes **Sprout** as "beginning to grow or sent forth shoots." **Piston**, "a reciprocating disk moved by pressure in a cylinder." **Cylinder**, "a solid generated by a rectangle turned on its centerline." **Stroke**, "a single movement that is repeated." **Pressure**, "the exertion of force." **Release**, "to let go," and finally much joy and happiness.)

"Thank you, honey. I really needed that!" Carl says as he tries to catch his breath.

"My pleasure, Carl. I enjoyed that almost as much as you did," Anita impishly stated while cleaning up. "Now you have thirty-five minutes to get to the airport."

"Oh shit!" Carl pulled up his pants and tucked his shirt in.

"It will take at least fifteen minutes to get his bags, that is if the airport is running as smooth as silk. And we know that will never happen…after all, it's New York," Anita said, trying to calm him down.

"Yeah, right! If I leave right this minute, I will be there by the time he makes it out of the terminal," Carl muttered to himself as he rushed through the apartment, gathering his shoes, wallet, and keys.

He gave Anita a kiss, followed by a big smile and a wink. "I will be back as soon as I can. If you head out before I get back, I will call you when I get in." Then out the door he flew, hopping on one foot at a time while he forced his shoes on.

At the airport, the PA system announced, "TWA Flight 117 from Cairo, now arriving at Gate 15." Tim shuffled off the plane, dodging people along the way. He started looking around for Carl immediately. He quickly made his way through the gateway to the main terminal area.

The atmosphere in the airport was loud and rude. Travelers are frustrated, trying to get to their connecting flights or trying to get out of the airport. Sarcastic remarks can be heard everywhere as the warm temperature and humidity fueled their mood.

Tim became angered because he expected Carl to be there to meet him. He lugged his bag toward the baggage claim area and exited to the street. Standing on the curb looking for Carl, he began to wonder if Haseeb was able to get hold of Carl.

Tim noticed the sharp sound of sirens in the distance and the hustle and bustle of the crowd. The warm, humid air saturated his clothes as sweat began to run down the back of his neck. He stood under the arrival sign, looking anxiously for Carl.

He was dressed in blue jeans and a white cotton shirt with long sleeves rolled up to his elbow with the roll tucked under the sleeve, not out. His head was adorned with his authentic *Indiana Jones* hat. With his white cotton socks and tattered tan, hiking boots, he stood out in any crowd. He had smooth shoulder-length hair, brown eyes, and stood 5'8" with a medium build.

Tim scanned the curb along the front of the airport; still no sign of Carl anywhere. He glanced at his watch and huffed. His flight had been on the ground for thirty minutes. "Shit! Where the hell is that asshole?"

Suddenly, he felt a hand on his shoulder. Thinking it was Carl, he blurted out, "I am going to kick your ass, Carl," and turned to face one of New York's finest smiling at him. "Good evening, sir! You realize that would be assault, right? What seems to be the problem?" the officer asked, still smiling.

Tim laughed nervously and explained, "I am so sorry, Officer. My ride is really late, and I was just blowing off some steam. I am fine now."

"I understand how frustrating it can be when you are in a hurry. Just try not to assault anyone while you are here in the Big Apple. I am sure your friend will be along soon. Sometimes the traffic can be unforgiving in this city."

"You are probably right, Officer! He should be here anytime now," Tim said as he continued to scan the traffic.

"Have a nice evening, sir!" the officer said as he turned to continue walking his beat.

"You do likewise, Officer." Tim noticed a silver car moving fast, swerving around traffic, coming into the arrival area. As the 1995 Mercedes-Benz came closer, he heard two horn blasts, and the car screeched to a halt in front of him. Carl had finally arrived.

Tim skipped all the polite salutations and jumped into the car before Carl could even open his door. "You are late, Carl!" Tim barked as he closed the door. "Let us get out of here and get straight over to the museum. I have got to get into the vault."

"Well! Hello to you too, Tim. I did not really expect a hug or a kiss, but a hello would be nice!" Carl said, and they pulled away from the curb.

"Hello! Now drive!"

"Sorry I am late, but Haseeb just called less than thirty minutes ago. I got over here as fast as I could," Carl fibbed as he sped off in the direction of the museum.

As Carl moved through the traffic, dodging cars that were going the speed limit, Tim explained, "As you know, Haseeb and I have been trying to piece together some peculiarities contained in some of the artifacts we have found over the years." Tim refreshed Carl's preoccupied memory.

"About nine years ago, we found this dig we are currently working on—we had no idea of its significance," Tim continued now that Carl's curiosity was piqued. Carl glanced over several times to check Tim's eyes. Each time Carl turned to look at Tim, he pulled the steering wheel to the left, forcing the car over the centerline into oncoming traffic.

Tim's attention was focused on the road as he was telling the story, and each time, he corrected Carl's mistake.

"Keep your eyes on the road, you idiot! Before you kill someone," Tim scolds.

"I am watching the road! Keep talking."

"The other night, I was comparing my logs against the scriptures, and I found an oddity that hit me like a ton of bricks. There were twelve original disciples with Jesus Christ. Their combined teachings comprised the oral gospel, which was put into script and eventually became the New Testament portion of the Bible. It is the fundamental teaching of Jesus as related by those that witnessed him."

"These twelve scrolls that we found at the dig are the scribed testimonies of Jesus Christ, which is a miracle. But then I discovered a thirteenth scroll. The thirteenth scroll contains, in my opinion, the last miracle to take place before the end of the world, as written in Revelations. It is the preamble for the second coming of Jesus Christ. I also believe it was written by John the Baptist."

There was an awkward silence in the vehicle. "No shit!" Carl shouted, focusing all his attention on Tim now. Carl's mouth dropped into his lap with his tongue dangling and once more allowed the vehicle to drift into oncoming traffic. Tim quickly grabbed the wheel to correct the danger and then chastised Carl for almost getting them killed.

Carl pulled himself back together and resumed control of the steering wheel. He also noticed flashing red lights in the rearview mirror and pulled to the side of the road.

The police officer pulled over behind him, exited his vehicle, and walked up to Carl's window, which was rolled down. "Good evening, sir! May I see your driver's license and vehicle registration please?"

Carl provided the documents requested and smiled. "Here you go, Officer!"

The officer looked over the papers for a moment then asked, "Do you know why I stopped you, sir?" He leaned in to hear Carl's reply and scanned the interior of the vehicle for any signs of alcohol or drugs.

"I think it was because I might have crossed the double yellow lines a couple of times," Carl confessed, looking at the officer to get his approval for being so honest.

"That is correct, sir. Have you been drinking today?" the officer asked as he put Carl's license into his top shirt pocket and opened the driver's side door.

"Would you please step out of the vehicle and come with me?" The officer escorted Carl to the back of the car, administered a field sobriety test on him, and continued to ask questions about alcohol or drugs in the vehicle.

Carl passed the test with flying colors, which puzzled Officer Gordon. "Sir, can you explain to me why you were swerving all over the road back there?"

There was a quiet pause for a few seconds while Carl formulated a reply. "Well, Officer, it's like this. My good friend here is suffering from altitude sickness. He just arrived from Cairo, and I am trying to get him to the museum doctor. He just cannot keep his hands off the steering wheel."

"Really! Well, sir, maybe I can help you a little." Officer Gordon walked over to the passenger side of the vehicle and asked Tim, "Sir? Would you please step out of the car?" The officer opened Tim's door and assisted him out of the front seat. "We cannot have you tugging on the steering wheel now, can we?"

With one hand on Tim's elbow and the other on the back-door handle, Officer Gordon directs Tim, "You are to remain in the backseat of this vehicle until your friend gets you to the doctor." He opened the door and assisted Tim into the backseat.

"Oh, Officer! I am afraid there is a misunderstanding here—"

The officer abruptly cut off Tim. "Sir? If you do not comply with my instructions, I will take both of you to the police station and have our doctor look at you. Do you understand me?"

Officer Gordan fastened Tim's seatbelt and closed the rear passenger door of the vehicle. "Everything will be fine, sir. Just relax and let your friend get you to the doctor. Okay?"

Carl desperately strained to contain his laughter as he saw Tim's face turning beet red with embarrassment. Officer Gordon returned to where Carl was standing. "Sir, under the circumstances, I am going to give you a warning this time. But please be more careful with your driving. You both could have been seriously hurt or even killed had I not stopped you." Officer Gordon handed Carl a warning and his driver's license. Then he returned to his patrol car.

"Thank you for your help! Officer, I am sure I will be able to take it from here," Carl responded as he opened the car door and entered. Carl fastened his seatbelt in silence and pulled into traffic slowly.

Once they turned the corner out of sight of the officer, Tim unloaded on Carl. "What the hell did you tell him? Are you fucking crazy?" Tim probed with fuming agitation.

"No! You are! Or at least that is what that officer thinks." Carl smirked and laughed out loud, unable to control it any longer. "I am sorry, Tim. I had to tell him something or he would have searched the car."

Tim, now puzzled, asked, "Why are you so concerned about the officer searching your car… Carl?"

"Well, let's just say I have not been a model citizen when it comes to parking in the city." Carl opened his glove box, and several tickets fell onto the seat and floor. He reached over and packed some of them back into the glove box and forced it shut.

"Shit! What the hell! Carl…are you starting your own display for the museum now?" Tim barked sarcastically.

"No! I just cannot seem to find the time to get them taken care of. But I assure you I will take care of them at the first opportunity. Now please continue with the story—it's absolutely fascinating." Carl continued driving toward the museum.

CHAPTER 7

Nickels in the Door

Tim continued with the story. "Because the thirteenth scroll has the last miracle identified in it, as told by John the Baptist, we have got to find it. After it is interpreted and confirmed with respect to its origin and content, we will be famous. Grants will come pouring in…we will get the best digs laid in our laps…we will be able to build our own museum in sunny California."

Carl, now excited about the potential, asked, "That sounds great, Tim! But whom are you going to get with the qualifications necessary for this kind of a find? You cannot just call up the Antique Road Show!" Carl knew that Tim had already contacted someone appropriately qualified but asked the question to impress Tim with his intuitiveness.

"You are absolutely right, Carl. It cannot be just anyone—it can only be someone the scientific community respects. That is why I have already wired Dr. Jessica Parker to come and verify the contents. She is the world's most renowned interpreter of ancient writings. Once she releases her professional opinion on the who, what, where, and when, we are home free." Tim turned to Carl, smiling from ear to ear, and stomped his feet on the floorboard of the car like a small child.

Carl demonstrated to Tim his ability to pay close attention to detail. "You forgot the why, Tim."

Tim looked at Carl with a puzzled look. "Excuse me?"

"The why? You forgot the why in your last sentence. You know! It is supposed to be who, what, where, when, and why. I am just saying you forgot it! What is up with that?" Carl asked, genuinely interested now.

"Everyone knows why! The end of mankind is why… Armageddon…fire and brimstone…the precursor to heaven on earth…that is why!" Tim looked at Carl, wondering if he had been paying any attention at all to his story. Thinking back on the interruption with the police officer, he realized he may have left that part out of the story.

A passage from a tablet Tim found many years ago consumed his thoughts now:

> *And in the final days a common man will bring forth a warning to all the world. He will be ridiculed and shunned by his peers. For his message will be one of despair and world catastrophes, the likes of which mankind has never seen. In the final hours there will unfold a breath of hope sung from the lips of a child who cannot speak. Perfect harmony from the voice of a child who hears not. And the world of man will cry, for they will know in their hearts the Lord's blessing is upon them.*

Tim leaned forward, over the from seat, and looked out the front window. The New York Museum of Natural History came into view.

A large Gothic building, majestic and serene, it was constructed in the 1940s. Forty white marble steps that stretched the length of the building led to the pillared archway and was blended with old stone and Greek marble. It was easy to understand why the building was considered one of the city's finest architectural landmarks.

There were two huge stone lions, one at each end of the lower steps, guarding the museum grounds. There were twenty gargoyles chiseled in marble, each perched on its own stone block, keeping a vigil watch over the museum entrance. Entry to the museum could be made through any one of eight doublewide glass doors. And there were eight identical doors located twelve feet inside the building with brass rails separating each entry.

Inside the museum was a maze of polished marble floors intermixed with oak floored showrooms and trimmed with polished colored stone. The elegant museum was divided by eras in history and covered as much about man as could be displayed. For the first-time visitor, the museum itself was breathtaking.

"This is one beautiful building," Tim shared aloud as he admired the museum.

"Yes, it is. I have been here fifteen years, and I am still captivated by its awe-inspiring design and engineering. They really knew what they were doing when they built this baby. She is an exhibit in her own right. She has hosted some of the finest exhibits in the world," Carl explained with deep admiration.

"If these stones could only talk, I bet they would have some unbelievable stories to tell," Tim speculated out loud.

"You know it, brother!" Carl agreed.

Carl pulled up in front of the shiny marble steps and parked the car. He got out and walked around to the steps and waited for Tim. Tim was still struggling with his door. Carl glanced over and started to laugh as he walked to Tim's door and opened it from the outside. "Child safety locks! The police officer must have tripped it when he put you in the back seat."

"Oh yeah! You are just a riot tonight, Carl."

Side by side, they hiked up the steps to the front door of the museum. As they got to the door, Carl ran his fingertips along the seam, where the two glass doors met, moving his hand up and down slowly as though he was searching for something.

"What are you doing, Carl?"

"I am trying to find the edge of my nickel. I put it in the seam of this door when I closed up for the night. I saw this trick in a movie once, and I thought it was a great idea. If the nickel is missing from the door seam, then someone has tampered with it," Carl explained.

"That is pure genius, Carl! What a waste of money those things are, hey!" Tim nodded toward the security cameras then looked back at Carl as though he had lost his mind. "Now! Come on, let us get that scroll."

"I know! To the scientific eye, this would appear to be ridiculous. But the security camera only tells me someone has been here after I am in the building. It does not tell me before I enter—this does."

"Okay! Okay! It's all good! Now come on and open the door already," Tim said anxiously.

"Not yet! I cannot find my nickel," Carl said, continuing to examine the seam of the door.

"Maybe the cleaning crew used the door for some reason and took the stupid nickel because you are not paying them enough," Tim responded impatiently, pressing his hands against the window to see if anyone was inside.

"No sir! The cleaning crew enters and exits the building through the rear door, where it is more secure. Besides, I have the only key for these front doors," Carl expressed while he pondered where the nickel went. "Oh! By the way, Tim, you cannot see past the inner doors because I had direct lighting installed on tracks between the inner and outer doors. It makes the inner doors act like one-way mirrors at night…you like that? That was my idea."

Carl put his face against the door window to check the floor for his nickel. Finally, he took his keys out of his pocket and put it in the door lock. As he turned the key and released the deadbolt, the door pushed toward them, slightly breaking the door seal. He heard his little security device hit the floor next to his left foot.

"It's about time!" Tim reached over to push the door open.

As Carl bent down to pick up his nickel, he got a strong whiff of rotten eggs. "Oh shit! Tim! Run like hell, it's gas!" Without saying another word, Tim bolted toward the marble steps. By the time Carl let go of the door and turned to run, Tim was at the steps.

Carl bolted too. "Sure! Asshole, you go first! Don't let my dead body get in your way!" Carl could hear Tim laughing as he caught up to him at the bottom of the steps.

"Who is the old man now? Pussy!" They both reach the last step at the same time.

Then…kaboom! First, a sudden thud and then the ground under their feet shook violently, followed by the blast that sent them both sailing over the hood of Carl's Mercedes. They landed on the other side of the car in the street. Shards of glass and other debris shot over their heads and into the vehicles and buildings across the street.

Shock from the blast knocked out the streetlamps, shattered windows down the entire block, and tripped car alarms throughout the block. Tim and Carl managed to pull themselves against the car and huddled together in a ball. They remained there semiconscious, listening to the airborne debris slam into the ground around them. Both looked at each other shocked, wondering what the hell just happened.

"Are you all right?" Carl asked Tim, genuinely concerned.

"Yeah, I think so. How about you?" Tim asked Carl.

"Sure, I am fine. Just a little dazed." Both stood up at the same time, brushing the dust and debris off their clothing and coughing to clear their lungs. They turned to view the damage to the museum and were surprised to see the damage was less than they imagined.

The museum was still standing a little less for wear. The glass doors to the entrance were gone, along with the framing that supported them. A black wire dangled from the corner where the security camera was. Inside the front of the museum, they could see several small fires and a lot of smoke. One of the marble gargoyles had come loose and mounted one of the lions. But the museum was still there.

"No doubt about it! I definitely paid the gas bill this month!" Carl shouted, trying to loosen some of the tension that could be felt by them both. "What do you think my chances are in getting my deposit back?"

Tim chuckled for a moment, still scanning the damage. Then he turned quickly and grabbed Carl by his collar. "Tell me you locked the shipment in the vault!"

"Hey! Easy on the threads dude—this is a $200 shirt," Carl responded, removing Tim's hand slowly. Tim took a few deep breaths and let them out, trying to relax a little.

"I'm sorry, Carl! You are right! I am just not myself right now. So help me out a little here."

Carl nodded. "Not a problem."

Then Tim grabbed him again around the collar and repeated his question. "Carl? Please tell me you locked the scroll in the vault."

"Hey! Hey! Hey! Chill, buddy! It's all good! Of course I locked the shipment in the vault. I do that with every new shipment we receive. It is standard policy…until we get authorization to display it. I'm sure its fine."

Releasing Carl's collar, Tim straightened it out and patted Carl on the shoulder. "I'm sorry, Carl! I have a lot riding on this one."

"No problem, Tim. I know your head's all screwed up in knots on this. Remember me? Carl? I have had to live through your miserable disappointments for fifteen years. I know exactly what is going through your mind right now."

"Oh yeah, buddy! I know! Right now, you are thinking about the Dead Sea scroll fiasco because you and Haseeb missed that discovery by two weeks. You are probably still blaming yourself on that one. But I say there is no way you could have known the boys had taken the Land Rover out to party that weekend. Hell! People run out of gas every day, man. It is just that they do not do it when they are about to discover history. No, sir, Tim, that was all on you, baby."

Tim could not tell if Carl was trying to reassure him or call him a loser! Carl could read the signs and tried to bring it home. "What I am trying to say, Tim, is it could have happened to anybody, not just you! So ease up on yourself. Listen, that scroll is inside a vault that is a foot thick. It is made of the toughest metal known to man on all six sides. Hell, Tim! It is impenetrable, and it is rated for temperatures that would scorch the sun. I seriously doubt that a little explosion could have even turned a page, let alone damaged anything!"

"Thank the Lord for that!" Tim started to feel reassured now. "I tell you, Carl, this has been one hell of a trip for me, and unfortunately, I believe it is only starting."

People began to gather around the area to see what happened. Sirens could be heard in the background now. They were rolling closer. Tim and Carl knew things were going to get a little hectic now.

They continued staring at the museum in shock. Carl tried to loosen some of the tension and blurted out, "I was thinking of moving the Egyptian exhibit to the front of the building anyway." Finally, Tim lightened up and started laughing. Carl joined him. They stood there surrounded in chaos and calamity, laughing the good laugh and happy to be alive.

CHAPTER 8

NYBP?

People flocked to the devastation like ants when their nest is disturbed. The street, once completely abandoned, was now alive with fervent activity. Emergency crews arrived. Firefighters donned their flame-retardant gear; other crew members stretched out hoses and connected them to the fire hydrants. Others with axe in hand and wearing their Scott SCBAs (self-contained breathing apparatus) entered the front of the flaming building to assess containment measures and to search for survivors.

Still others positioned themselves around the building with hoses and ladders, dumping tons of water into the building to douse the flames. Tim looked directly at Carl without saying a word. Carl immediately recited, "Watertight to fifteen thousand pounds per square inch. It's tighter than a B-Class submarine." Tim just smiled then turned back to the action.

One group of firemen climbed into what looked like a fancy cherry picker basket, and they were raised to the roof. They stepped from the basket onto the roof and disappeared.

Police officers ran back and forth around the scene looking for witnesses and set up traffic barriers to control traffic. They had blocked off the streets around the museum and set out yellow caution tape to block the flow of curiosity seekers.

EMT personnel stood ready to provide emergency care to any victims found in the building. Ambulance personnel were on scene too just in case they needed to get victims to the hospital. But so far, that was not the case.

The once empty street was filled now with hundreds of spectators shuffling through the debris and emergency crews to get a closer look at the damage to the museum. Police had their hands full pushing them back. It reminded Tim of the carnival with so much activity going on. Carl thought to himself, *How funny, all these people crowding to catch a glimpse of the museum. Where was this crowd when the King Tut exhibition was here on display for twelve weeks?*

Smoke billowed from the main entrance of the museum as the firemen doused the smaller fires just inside. The smoke lay low and heavy in the windless night sky, giving the illusion of a grayish cloud cover over the city. The confusing din was effective in smothering the commands yelled by police and firemen. Bull horns and mobile public address systems had to be used for communications to the groups of emergency workers.

Tim and Carl were being treated for minor cuts and bruises by emergency medical technicians. They sat on canvas folding stools, observing the mass calamity, while EMTs checked their vital signs. They were given a clean bill of health.

They stood watching, dazed, and confused witnessing the laudatory efforts of the local emergency teams, admiring the precision and accuracy of the groups working together, completing one task after another without a single word spoken between them. One could only be inspired observing the synchronously orchestrated event, people of different ethnic backgrounds and various cultures working in unity to achieve a common goal.

Several moments passed as Tim and Carl stood witnessing this marvelous event. Tim's thoughts drifted to the safety of the shipment in the vault, while Carl's thoughts were more about the insurance coverage of the exhibits that were surely damaged by the explosion. Carl also wondered how long it would take to clean up the mess and how soon he would be able to reopen the museum.

"Excuse me?" A voice echoed from behind them.

They turned around to see an officer there. He was mighty young to be a police officer. He looked like he had just graduated high school and was possibly eighteen years of age. The officer's hair was shoulder length and in contrast to all the other officers out there. He was thin and lanky with discernibly dark eyes. The kid was odd-looking and out of character for New York's finest. But looking at him, you could not pinpoint a specific peculiarity that was over another.

"Excuse me! I am Officer Damon, and I need to ask you a few questions for the record. I need to get a statement about what happened here tonight. Do you have a few moments?"

"Sure, fire away, Officer," Tim replied, eager to help.

"First of all, I need to get your names and addresses so we can contact you later if necessary."

"Well, I am Tim Wilford, and I just arrived in town tonight from Egypt. I really do not have a place to stay yet, but I was thinking that Carl here would put me up at his place until I decide what I am going to do for accommodations," Tim said as he looked at Carl to see his mouth give a little uncomfortable twitch.

Carl looked back at Tim with a shocked expression on his face. "You have not made any arrangements yet?" Carl asked him to seem surprised. "What if I have plans that do not include you and involve my being alone in my suite? Did you ever think of that, or did you just think that Carl would put you up, so why concern yourself with such trivial pursuits like finding a place to stay?" Carl said apathetically.

"You gentlemen can discuss this later. Right now, all I need is an address where you can both be reached," the officer interjected.

"Fine! You can reach us at the Canterbury Penthouse Suite. If Mr. Wilford is not with me, at least I will know where he can be reached," Carl stipulated.

"And you are?" Officer Damon asked, looking at Carl.

"Carl Curry, with a C."

"Now can you tell me what happened here tonight? Beginning with when you arrived and everything up until the explosion." The officer prepared to write down all the information.

"Well, we arrived about an hour ago to check on Tim's shipment. As I unlocked the front door, I bent down to pick up the nickel I put in the door as a first-line indicator of anyone tampering with it when I am gone. Then I smelt gas and yelled at Tim, and we ran down the steps. When we got to the last step, we were blown over the car and landed in the street. We were unconscious for a little bit, and when we came to, we heard sirens. The emergency units showed up. The EMTs checked us out. And now you are asking us questions. That is about it!" Carl summed it up quickly, not trusting this person in a policeman's uniform.

"And do you affirm this statement, Mr. Wilford? Or is there anything else you would like to add?"

"Yeah, that is pretty much the whole story, Officer," Tim agrees.

"Now let us get back to this shipment you referred to earlier in your statement, Mr. Curry. What exactly was so important about this shipment that you had to come here this late in the evening?" the officer inquired.

"Well, you see—" Tim began to explain but was quickly interrupted by Carl.

Carl had been trying to figure out why his senses were on alert and why this officer was different than the other officers on the New York Police force. Finally, it hit him like a ton of bricks. The badge on his uniform had NYBD on it instead of NYPD like the rest of the officers. This guy was a fake!

"You are no policeman. Who the hell are you?" Carl challenged. Carl called to the officers in the area. "Is this guy with you? Hey, Sergeant! Is this guy with you?" Carl singled out a uniformed officer with three stripes on his sleeve.

The sergeant glanced around to look directly at Carl. "What is it? Can't you see I'm busy here." He flared his arms about the maddening scene.

"Is this guy one of your officers?" Carl pointed to where the fake policeman stood.

"What guy?" the sergeant quizzed impatiently.

Carl turned around to see that the fake police officer who had been questioning them had vanished. He quickly looked in every direction to see if he could find the faker, but he had disappeared out of sight.

"Never mind, sorry!" Carl yelled back to the frustrated sergeant as he shrugged.

"Okay! Where the hell did that little shit go?" Carl asked Tim, who stood dazed, looking into the distance with his eyes wide open.

"I don't know! I turned my head to see whom you were yelling at, for one second and when I turned back around, the guy was gone. I mean he had disappeared…vanished…*poof*, he was gone as though he was never here. This is freaky, Carl! And I do not like it at all. It gives me the willies!"

"Don't sweat the small shit, Tim! We got other stuff to worry about. No sense wasting time worrying about a speed freak," Carl said, trying to get Tim's mind refocused back on the scroll. "Besides, we will probably see him again. He knows where we live now." Tim turned to Carl and nodded in uncomfortable agreement.

Tim and Carl turned their attention back to the emergency crews and the museum. It would be some time before they would be allowed to enter the building. And the real police still had to get information from them about what happened.

*****

Meanwhile, somewhere in the desert mountains of New Mexico, a sacred ritual was being performed by an Indian shaman. The spiritually guided shaman chanted sacred passages from his ancestral past. The fire spiked into the desert night sky, causing the surrounding barren landscape to illuminate in a bath of light. Clad in his ceremonial attire of buckskin chaps and a coyote pelt covered with eagle feathers and a bear's tooth necklace, he shook his talking stick.

The talking stick was wrapped in soft hide with strings of beads and feathers. It had a coyote scull attached to one end and a tortoise shell tattle on the other. It was used to drive away evil spirits and assisted in calling to the spirit world. Guided by visions, he called upon the Great Spirit to send forth a great warrior spirit to protect the unsuspecting white man who went by the name of Tim Wilford.

The shaman called Soaring Eagle watched the flames form the face of the Great Warrior Spirit. The face that was formed by the flames of the fire in the desert's night sky was that of the Great Warrior Geronimo. Soaring Eagle kneeled before the flames with his palms and face pointed to the sky, thanking the Great Spirit for granting his request. Soaring Eagle remained there for several days in deep prayer, giving thanks to his ancestral spirits for their guidance.

*****

Back in New York, Tim and Carl had finished giving their statements to the police. The building was now being aired out by large, commercial-sized portable fans. The firemen were rolling up their hoses and putting their equipment away.

Carl was hoping that the explosion's bark was going to prove worse than its bite. But at this moment, he could only think the worst. There was a lot of work left to do in assessing the damage and notifying the appropriate parties. This is not going to be a good day, Carl thought.

Tim was anxious to get in the building and check on the shipment. This was his only purpose and concern. He was determined to see this thing through, now more than ever, to the end. Obviously, Tim thought, he was not the only one that wanted that scroll. He was going to have to be incredibly careful.

*****

Sitting high above the city on a tall building's ledge across from the museum, a loathsome figure rousted, devilishly contemplating his nest move in this seemingly endless game. With his large, clawed hand, he hoisted Damon's severed head to eye level. Damon's eyes and mouth were still wide open with the expression of terrifying fear. The creature gazed at the grimacing face with pride and spoke. "What to do! What to do! What to do!"

His powerful voice rumbled. "What is the matter, cat got your tongue?" He gleefully admired the head of his victim. "Oh! I see you are shy or perhaps you cannot think of any more excuses! Well, most certainly you have no more excuses." He chuckled softly. He placed the top of his victim's head into the palm of his right hand and with no more effort that it took to squash a tomato, he crushed it. The sound of crunching bones and squishing flesh caused him to shudder in sardonic ecstasy.

CHAPTER 9

Who Has the Wishbone?

We jump forward in time a little to a parking lot in the city of New York where some interesting twists are about to unfold, just a few weeks before Christmas.

Perched on the ledge of a tall building in the distance, you could just make out a dark, lurking figure surveying the events in the city from his aerial vantagepoint.

His attention was drawn to a desolate parking lot in the city. He seemed to be waiting for something to happen. As we drew nearer to the figure, it was the same menacing creature that had crushed Damon's skull in his hands. He appeared to be much more amused tonight.

In an empty store parking lot, two gang members were conversing loudly about where their drug runners were. "They are late!" said Crusher, the bigger of the two seemingly upset. He was partly disgruntled because of the weather and mostly because he did not like to be kept waiting. He was the leader of the gang and expected his people to be on time with no excuses.

Snow was falling steady and heavy. There was about four inches of new snow on the ground. The night was dark, and the only light they had were the lights from the car. It was still running, leaving a constant cloud of smoke and steam trailing from the tailpipe. There were thinly scattered light poles around the parking lot. But they offered little help because of the thick snow. The scene was quite desolate, and the leader was getting nervous.

A dark car rolled into the other end of the parking lot, heading toward the store at first then turning slowly toward the two men. It stopped for a moment, and then a red flashing light appeared in the front window. The car slowly rolled toward the two men, crushing the soft snow under its tires and leaving a clean trail of tire prints.

Crusher shifted his anger from his late runners to the current situation. He told his lieutenant, Hambone, "Relax! There is no reason to panic. We are clean. All the stuff is with the runners. Go to the front of the car and put your hands on the hood. We will play it out with the pigs. This will save some time, and when the cops leave, we can get down to business." Hambone and Crusher walked to the front of the car, turned, and put their hands on the warm hood of the car.

A tall man with an ankle-length, well-worn, and partially torn trench coat entered the parking lot from another angle. His view of the situation was perfect. As he slowly approached, the car with the flashing red lights on the dashboard sped up to the other car, headlights to headlights, until it was just twenty feet away and then stopped. The doors swung open, and two men jumped out, positioning themselves behind the open car doors. Both men had shotguns pointed in the direction of the two gang members who had their backs turned to them.

A burly voice yelled out, "Assume the position, assholes!"

Hambone turned his head slowly toward Crusher as they both spread their legs while facing their own vehicle. "Crusher? That ain't no cop! that is Wishbone!"

Crusher's eyes opened wide, then he bowed his head toward the car as though he was thinking, *Shit, shit, shit!* He looked back at Hambone and whispered, "We are so fucking dead! We got to split, man… We go on three. One…two…"

Three was interrupted by two loud bursts of gunfire echoing into the still of the night air. Both shots hit their marks. Hambone and Crusher slumped over the hood of their car, coughing and spitting up blood. They looked into each other's eyes as their bodies started to slide off the hood. Two more blasts from the shotguns finished the job. The car radiator started to steam, pouring hot water onto the red snow.

Crusher and Hambone's bodies fell lifeless into the fresh snow in front of their car. The blood oozed from their fatal wounds, forming a large pool of red snow around their bodies.

The ghoulish figure danced an ominous jig atop the ledge of a nearby building. He unleashed a horrific, ear-piercing screech that shook the snow out of the trees for miles around the building.

Wishbone looked over at Horse, his sidekick, and asked, "What the fuck was that?" They both looked around, trying to pinpoint where that godawful cry came from.

"I have no fucking idea. Let's get out of here!" Horse responded, climbing into the passenger side of their car.

Job complete, Wishbone and Horse exchanged childish dialogue. "Timing is everything, my dear Horse. They had no clue what was coming. It was perfect!" Wishbone shared with his partner.

"It was fucking brilliant, boss!" Horse replied as he turned off the flashing red light on the dashboard and put it on the floor. As Wishbone put the car in gear and started to drive away, Horse offered his parting remarks. "Night-night, tough guy! Sweet dreams in never-never land."

"What do you want to do about the old bum over there, Wishbone?" Horse asked, looking toward a tall figure walking toward them.

"Drop him on the way out my friend! No witnesses!" Horse loaded two rounds into one of the shotguns and rolled down his window. Wishbone pulled the car around and headed directly for the old man.

By this time, the old guy had traveled half the distance between the scene and the other edge of the abandoned parking lot. He moved smoothly and surely, as if he had a purpose for being there. He saw the car coming at him and didn't even flinch. He seemed to ignore the car and its occupants. As the car approached him, it slowed down. The old man looked directly at the man with the gun and stopped, as though he wanted Horse to have a perfect shot.

Again, the silent night air was filled with a thundering sound "No! No! No! You fool! Leave now. He's not what you think he is. Leave now!" An eerie, thundering voice bellowed across the parking lot, vibrating the ground and causing more snow to fall from the branches of trees.

The old man turned and looked directly at the dark, menacing figure perched on the ledge of a building four blocks away. He looked directly into the night creature's eyes. It was as though they were standing face to face. Satan's choice servant smiled and said, "Forgive them for they know not what they do?" then laughed and leaped a hundred feet to the next building. Moving closer while taunting the old man with his gestures, he added, "Thou shalt not kill."

Watching now, from his new perch, he knew there was nothing more he could do. He started humming a satanic chant that moved swiftly through the night air like a nimrod straight into the ears of the two men in the car. It intensified their rage toward the old man. The old man snapped his fingers, and the building that the demon was perched upon whipped violently, launching the creature twenty-seven floors down to the concrete below. As he fell, he heard the old man's voice: "Vengeance is mine, saith the Lord thy God!"

The beast crashed deep into the concrete sidewalk. Crushed from the impact, his body was contorted in a mass of broken bones and mangled flesh. Slowly, his body was dragged into the earth below and disappears. Now with the creature dead, the two youths had no one to reverse Satan's spell of rage.

Horse laughs at the old man, "Ha! Ha! Ha! You made a big mistake tonight old man; you should have stayed home in your box!" The car was within several feet of him, and Horse fired the shotgun. The slug passed right through the old man's chest and sent a puff of snow up quite a distance behind him as it struck the ground. The old fellow didn't budge an inch. Horse could not believe he had missed him at this range. Wishbone yelled at Horse, "Hurry up, asshole, we have a lot of paybacks to dish out tonight. We can't be fucking around with this guy all night."

Horse jumped out of the vehicle as Wishbone slowed the car down in front of the old man. Wishbone calmly looked at the old man. "Kids today! What are you going to do?" He shrugged and smiled. "He spent his anger management money on drugs. Don't worry though, this isn't personal, it's just business."

Horse moved to within five feet of the old man and yelled, "Nighty night!" and pulled off another round dead center in his chest. The shot passed through the old man again and didn't even move the threads of his clothing.

Wishbone yelled to Horse, "Get in the fucking car, dude! Let's go!" Horse jumped in, and Wishbone stepped on the gas, spun the car around, and ran to the end of the parking lot where the two bodies were. He lined the car up and stepped on the gas, heading straight for the old man. "You can't miss with a fucking car, baby!" Both men started laughing, hoping to see the old man's face as the car smashed into his body.

The old man shook his head, wishing the two men would drive away. Now they left him with no choice. He stretched his hands straight out to his side and faced the car with the two men in it. They reached about forty-five miles an hour before contacting the old man. The old man did not move; loud scraping and screeching of metal filled the air with white, yellow, and orange sparks shooting into the darkness of night. He passed right through the vehicle. With his hands open and extended, he removed both of their hearts as he continued through the back of the vehicle, splitting it in two pieces.

Wishbone and Horse's heads fell forward and silent; they had two large holes in their chest's where their hearts used to be. The two pieces of the car continued to travel forward on their own as they slowly lost speed and then lay over in the snow, sliding to a stop against the curb. Their bodies lay lifeless in their respective sides of the car.

The old man dropped the two still beating hearts into the snow and continued to walk toward the other bodies of Crusher and Hambone near the front of their vehicle. As he approached the area, the spirit images of both men rose out of their bodies. They looked confused at each other and then looked toward the old man. "What is going on, man? Are we dead?" Crusher asked. The old man did not respond but looked around as though he was waiting for something.

Hambone screamed, "No! Please No!" and pointed to his own body lying on the ground, lifeless in his own blood. "Oh, fuck! We're dead, Crusher. Look!"

Crusher looked down at his own dead body and turned away in disgust. "What the hell! My guts are all over the place. Oh please, don't let my mother see this—it'll kill her for sure." The old man stood silent.

The winter silence was shattered with unearthly cries and screams now coming from all around them. A rumbling noise broke through the shrieks. Crusher and Hambone attempted to grab each other for support, scared, confused, and unsure of what the rules were in this game. Hambone looked to Crusher for answers, but he didn't have any. Hideous shadows were seen moving on the snow, closing in on them in a tight circle.

Casting those ominous shadows were demon soldiers of souls, gruesome monsters from hell, whose only purpose was to collect the souls of the damned. They stood only three to four feet tall but cast enormous shadows that were three times their size. The demon soldiers themselves weren't much to look at; they looked like the Pillsbury Doughboy on steroids after he'd been left in the oven for six hours, charred flakes falling off into the snow. The shadows of these soldiers were the mean ones; they consumed souls. They inflicted unspeakable pain and horror on new souls as they dragged the souls kicking and screaming down to hell and presented them to Lucifer.

The clean, fresh snow turned pitch black in their path as they approached the young men. The sounds of recently captured souls screaming in agony and terror put fear into the men's hearts. They dropped to their knees together as the circle tightened around them and they cried out, "Oh, God, please forgive us. We didn't know! We didn't know!"

The circle was only a few feet from the men, and the shadows were so tight you couldn't distinguish one from the other. Crusher and Hambone looked helplessly at each other as they witnessed their hands being consumed by the shadows. They both screamed out in pain as the process of change began. They desperately looked at the old man in the trench coat as he stood witness to the event. They pleaded with him one last time, "Please, mister, help us! We'll do anything."

The old man looked at them struggling to get free, then he heard them start to recite the Lord's Prayer. "Our Father…deliver us from evil…" He raised his right hand with his palm facing the two young men. A small white circle was glowing in his palm.

Crusher and Hambone felt a funny sensation overcoming them and looked at each other in disbelief. They yelled out "Wow!" as though they were topping the crest of an enormous roller-coaster ride.

The shadowy figures squirmed violently and screeched out disdainfully. The toasted demons showed their long snouts and sharp teeth to the old man and hissed their objections at him.

The old man closes his fist, securing the two spirits in his hand. He ignored the awful, horrific noises of the demons and waves his left hand in a half circle over his head. The evil creatures were powerless against the old man and started to move away reluctantly.

There was a clattering of voices at the other end of the parking lot where Wishbone and Horse had realized that they were dead. They started yelling obscenities at the old man. The demon soldiers hesitated to see what the old man would do. He turned his back on them, and the horde of terror marched toward the other two men.

Along the way, a couple of them started playing kickball with one of the hearts left on the ground. When the heart went airborne, one of the creatures reached out and grabbed it. He sniffed it and then took a bite out of it and spit it out on the snow.

The old man could hear Wishbone and Horse inviting the creatures over to get a piece of this. Horse yelled out to them, "Come on, bitch, you want some of this? I got something for you." Seconds later, the two men were screaming in agony. Spewing blasphemy and cursing, they cried out into the dead of night. Moments later, the screaming disappeared, and the demons departed with their bounty of souls.

Once again, the night was quiet and calm. The old man pointed his left fist at the two lifeless bodies in front of the car. He opened his hand to expose his right palm, and out shot the spirits of Crusher and Hambone, pouring out like the burst of a double-barreled water gun. Both were cast back into their bodies.

The bodies shuddered and shook from the impact of their spirits rejoining their bodies. The old man waved his right hand in an arch over his head. Crusher and Hambone came back to life, healed of their previous wounds. The blood on the snow disappeared as it flowed back into their bodies.

Crusher and Hambone poked at each other to see if they each were alive. They hugged each other, thankful to be back in their bodies and equally thankful that the demons were gone. "Who are you, old man?" Crusher asked.

"They call me many names, but you may call me John. Your salvation is at hand, my friends, and your sins have been forgiven. There is one task that remains for both of you. It will be revealed to you tonight."

Crusher asked again, "What are you, John? An angel?"

John looked at Crusher and called him by his real name. "James Allen Johnson, who do you believe I am?"

Horse couldn't suppress his desire to blurt out, "Holy sh—" His voice was stopped, but his lips kept moving to complete the sentence.

John held his two fingers together and expressed to Horse, "Carlson James McCall, do not spew words that are sacred out like a mouth full of hot coffee!"

Horse's lips were released, and he apologized. "I'm sorry!"

Looking back at Crusher for a reply, John asked again, "Who do you believe I am?"

Crusher looked into John's eyes then his long, straw-like brown hair and his ungroomed beard. He put his hand to his chin and scratched it in thought, then while focusing on his dark skin, the mark around his neck, he snapped his fingers and spoke. "I believe you are John the Baptist, sent by the Lord to prepare the way for his son, Jesus."

John paused for a moment then replied, "You are well-studied in scripture! I'm curious as to why you take the gift of life so lightly. You sell the poppy product that harms so many of God's children for gold and power. Why?"

James Johnson looked at his feet, ashamed of his life choices and what he'd done with it. Carlson McCall also looked down, ashamed of what he had done.

John felt their remorse and raised their spirits. "Fear not, my friends! The Lord indeed loves you both. Put aside these childish things and walk tall, be humbled. The Lord has a task for you both." James and Carlson raised their heads and looked into John's eyes.

"Hasten yourselves to the task at hand, my friends, for the time of a great miracle grows near. You have one week to complete your mission."

James and Carlson looked confused. "What mission is that, John?"

"It shall be revealed to you in time. May the Lord bless and keep you on your journey, amen!" As he completed his instruction, he turned and walked back the way he came, leaving no footprints in the snow.

As the old man disappeared, James and Carlson looked at each other and embraced. "Wow! What a night!" Carlson remarked as they approached the car.

"Yeah! No one is ever going to believe this night!" They both glanced across the parking lot to see two halves of a car, each on its side and steaming in the snow.

Carlson asked, "What's that over there?"

James replied as he got into the car. "I don't know, and I don't care. It's time to get out of here."

James started the car and drove a few feet. The car coughed and chugged then stalled. Looking at the gauges, he said, "I don't believe this! We ran out of gas!" They looked at each other dumbfounded and then broke out into laughter.

CHAPTER 10

Lost and Found!

Robert Cummings, a twenty-year sergeant with the New York City Police Department, commented to his partner, George Thorpe, "The violence in the city has never been greater than it is today. Patrol cars travel in packs of two or three, with two or three officers in each car…just to stay alive."

Officer Thorpe replied, "Yah! You're right! It's totally insane out there."

The radio wailed with a dispatcher's voice, calling for all available units to respond to a parking lot North of Grand Avenue and Sixty-Sixth Street. Sergeant Cummings picked up his handset and acknowledged the call, while Officer Thorpe switched on the lights and siren. They sped up to Sixty-Sixth Street and headed north, maneuvering around double-parked vehicles and avoiding cars failing to yield the right of way. With fewer patrol cars and officers to respond to traffic infractions, this had become a major problem for police and other emergency response vehicles. Sergeant Cummings and Officer Thorpe simply dealt with it.

Officer Thorpe called the dispatcher on the radio and advised they would be on scene in approximately three minutes. He asked for more details on the call. A few seconds passed, while the dispatcher gathered the information. Then the dispatcher announced, "Be advised, shots fired…no further information available at this time." Officer Thorpe readied the shotguns mounted in the center of the patrol car.

As they pulled into the abandoned store parking lot, they noticed a second patrol car rolling into the parking lock from the opposite side. Sergeant Cummings's car entered from the north and the second car from the south. The vehicles slid side to side as they cruised through the snow-covered area. Officer Thorpe could barely make out the silhouette of a vehicle with its lights on at the east end of the parking lot. Neither of the two patrol cars could determine if anyone was in the car.

The snow was coming down heavy, thick as feathers from a pillow fight. Visibility was only about fifty feet. Officer Thorpe radioed over to the other patrol car and advised them to pull to the rear of the vehicle, and they would approach from the front to prevent an unwelcome attempt to depart. As the patrol vehicles rolled into position, they stopped approximately ten feet from the front and rear of the suspect vehicle.

With shotguns drawn and ready, Sergeant Cummings and Officer Thorpe exited their vehicle, while the other officers stayed in their unit. Sergeant Cummings grabbed the handset and switched it to megaphone. "Driver of the vehicle, this is the NYPD. Step out of the vehicle with your hands above your head." A few seconds passed, and Sergeant Cummings repeated the order. Still no response. Cautiously they approached the car on both sides. They discovered that the vehicle was abandoned.

Sergeant Cummings reached through the driver's door to grab the keys from the ignition. A passing car backfired, sending Officer Thorpe down on one knee. Sergeant Cummings lost his balance in the new snow. The other two officers jumped out of their car and drew down on the vehicle, ready to open fire. Sergeant Cummings had slipped right on his ass next to the vehicle; with his hands still inside the driver's window, he tried to pick himself back up. Still sliding on the slippery snow, he looked like a clumsy ballet dancer.

Officer Thorpe, realizing there was no threat, yelled, "Clear!" The other officers holstered their weapons. He headed over to Sergeant Cummings to help him up.

Sergeant Cummings finally got a foothold and jumped up quickly and brushed the snow off his uniform. He quickly turned to Officer Thorpe and said preemptively, "Not a word! Not one lousy word!"

Officer Thorpe giggled to himself and turned to look at the car more closely. He noticed steam was coming out from under the hood of the car, indicating it hadn't been parked there very long and was recently abandoned. "Sergeant Cummings! They can't be too far from here—the car is still warm."

Officer Thorpe stated, "I'll call in for a tow truck," and headed for the squad car. Sergeant Cummings acknowledged him with a nod then turned to Officer Bob Buttes and his partner, Officer Tom Rush; both were laughing by their unit. Sergeant Cummings stood directly in their headlights and flipped them off with both hands while showing a smiling smirk.

Sergeant Cummings walked to the other side of the suspects' vehicle as Officer Buttes and Officer Rush climbed back into their patrol car. He waved his flashlight on the ground, looking to see if there were any tracks. At the front of the car, he noticed two sets of footprints, slightly covered with new snow, heading toward the east wall of the parking lot. Looking toward the brick fence along the east side, he noticed the trail disappeared over the wall heading toward Sixty-Sixth Street.

"Officer Buttes! You and Officer Rush want to cruise on Sixty-Sixth Street? I got two fleeing over the wall, two sets of tracks."

Officer Buttes yelled back, "You got it, buddy. We'll check it out. If nothing turns up, we'll see you guys over at Darnel's Donuts. Coffee's on us!" He slowly pulled away.

Sergeant Cummings nodded to them. "Thanks, asshole. Coffee's been free at Darnel's for twenty-five years."

Sergeant Cummings focused on the vehicle and completed an unrevealing search. He walked to the front of the car and noticed several gunshot holes in the radiator and scratches on the hood. The snow in front of the headlights had been disturbed as though someone was sitting or lying down in the pile of fresh snow. Two imprints of bodies lying in the snow in a fetal position was very interesting.

He stood quietly examining the scene and then looked toward their patrol car. He saw another set of fresh tire tracks that were made recently. As his eyes begin to trace the tire marks, he noticed another set of footprints. "Officer Thorpe!" Sergeant Cummings yelled out to be heard over the idling engine of their patrol car. "You should see this!!"

Officer Thorpe got out of the patrol car and almost fell on his ass. He checked the ground for whatever it was he stepped on and noticed a 12-gauge shotgun shell casing. "Oh yah! Shots were definitely fired here tonight, Sergeant Cummings!" He reached down to grab the shell casing and saw another just a few feet away. Officer Thorpe walked over to where his partner is standing, carrying the results from his plate check and the shell casing.

"This vehicle is registered to James A. Johnson, a.k.a. Crusher," Officer Thorpe shared as he presents the shotgun shell casing to Sergeant Cummings. "Look at this—Crusher's vehicle has several holes in the radiator. Either someone is a lousy shot or we're going to find a few dead stiffs tonight."

"What's that?" Sergeant Cummings asked as he took the shotgun shell casing from his partner.

"Looks like a twelve-gauge slug casing, I believe," Officer Thorpe offered as he handed the casing over to Sergeant Cummings. "That must be what made the holes in the radiator. Why aren't there any bodies?" Sergeant Cummings shared out loud. "You'd have to be blind, not to hit your target from ten feet with a slug this size."

"Sergeant Cummings? There's something else you should know. There's a set of car tracks over by our car and they don't belong to us or Officer Buttes." Officer Thorpe pointed to a set of tracks parallel to there about six to eight feet from their car. Sergeant Cummings scanned the tracks with his flashlight, following them as they pulled away from the scene and headed toward the center of the parking lot. Then they spun around and came all the way back to the curb behind the suspect's car.

Sergeant Robert Cummings walks to the back of the suspects' vehicle and notices mud and snow mixed and thrown back ten to fifteen feet and the tracks shoot to the center of the parking lot. Not much farther away than the original tracks made before. Sergeant Cummings rubbed his eyes as though he couldn't believe what he was seeing. At the end of his flashlight beam, the tracks of the other vehicle seemed to separate, traveling in separate directions.

He turned to Officer Thorpe. "Move the car out of the parking lot to the street. Do it carefully. Ty to stay inside your own tracks as you back it out. I think we have another problem." Officer Thorpe complied with Sergeant Cummings's instructions and backed the car out of the parking lot the same way they pulled in. He got out of the squad car and stood next to it while watching his partners silhouette in the falling snow. His partner was moving slowly toward the center of the parking lot. His flashlight was shining on the ground and moving from left to right.

At the center of the parking lot, Sergeant Cummings bent down and examined an object in the snow. "Holy shit! What the hell is that?" He poked at a roundish bloody object in the snow that appeared to be an organ.

Suddenly, he jumped up gagging, stumbled, and fell backward in the snow and yells "That's a fucking heart!" Officer Thorpe, seeing his partner fall, hustled toward him to help.

"You all right, Sergeant? What's the matter?" he yelled out to his partner as he covered the distance in seconds.

"I'm all right, George! Just a little sick to my stomach." Officer Thorpe saw what had his partner gagging and heaved his cookies up in the snow. "What the fuck is that? It looks like somebody's heart with a bite taken out of it... Gross!"

As he composes himself, Sergeant Cummings replied, "That's exactly what it is, I've never seen anything like this before."

Sergeant Cummings stood up and continued to look around with his flashlight and saw another heart on the ground close by. "Now I'm sure we're going to find some stiffs tonight. Just not with all their pieces." He continued to examine the scene and saw that the car's tire tracks started to separate apart. "How the hell is that possible? These tracks are perfectly even all the way to this point here and then they start to widen."

"Motorcycles? Maybe!" Officer Thorpe offered as a solution to the puzzle.

"No! Can't be motorcycles because the tracks are perfectly parallel to this point." They continued to follow the tracks another thirty feet, and then they began to run toward the end of the tracks they were following. Officer Thorpe got to the end of his track first and yelled to Sergeant Cummings, who was reaching the end of his track about fifty feet away.

"You're right, Sergeant Cummings. They aren't motorcycle tracks. The car was split in half. I got my half over here."

"Same here, half a car and a dead boy at the wheel…looks like a gunshot wound to the chest," Sergeant Cummings replied.

Officer Thorpe checked his half of the vehicle and found the passenger with a hole in his chest. "That's what I have here, same wound to the chest." Sergeant Cummings pushed the button on his portable radio.

"Dispatch! Call Captain Belmont and tell him we have a double homicide on the eastside of the Foster Farm warehouse parking lot on Grand Avenue and Sixty-Sixth Street. We're going to need a forensic team and homicide investigation team as well as the coroner out here. We have two young deceased males on scene. Apparent gunshot victims." Sergeant Cummings finished his transmission.

"Officer Thorpe? Go tape off the entrance to the parking lot and run the plates on this vehicle. I want to know who we have here."

Officer Thorpe looked at his victim and then walked over to Sergeant Cummings and glanced at the other victim. "No need to run the plates—I know who they are. This one is Wishbone, leader of the Warf-rats, a big group that runs guns and drugs into the city. They used to own the docks for several years," Officer Thorpe rattled off as he examined the face of the driver more closely.

"The other guy is his top henchman, Horse. He's been with him for about seven years. Together they were unstoppable. That is until Crusher and Hambone popped up in the city and took over the Dark Angels. They put a stop to the Warf-rats' activity in the city and held it. Until now, word on the street was they had hammered out a treaty that gave the Warf-rats control of a five-mile block along the docks. The Dark Angels kept control of the inner city. Both groups are no good troublemakers and wouldn't give a thought to blowing away a child or a cop."

Sergeant Cummings looked at his usually short-sentenced partner with amazement. "Where did you learn about these guys?"

Officer Thorpe looked at Sergeant Cummings as though he may have forgotten to mention to him a thing or two about his past. "Port Authority for seven years before I made the switch over to the municipal side of the house. Sorry I didn't tell you before, but the matter never came up."

Sergeant Cummings shook it off and directed Officer Thorpe over to the patrol car. "Let's get the tape up and hold that tow truck off until the investigation team gets out here. We don't want anyone messing around out here until they clear it." After tying off the last piece of tape, Sergeant Cummings and Officer Thorpe sat down in their squad car and started filling out their reports. After ten minutes, the first of several squad cars rolled onto the scene, carrying the captain.

"Sergeant Cummings, what do you have here?" Sergeant Cummings jumped out of the car and started to brief Captain Belmont on what they discovered from the time they got the call until they discovered the bodies.

Officer Thorpe got excited when he saw the forensic mobile unit roll past. "You know that lab can test any material in the field and produce immediate results." He shared another long sentence with Sergeant Cummings.

Robert looked over at George and said, "Don't tell me. Let me guess—you worked in the forensic division of the Port Authority for five to seven years."

"No, Robert. I saw a special on public television concerning the timely processing of forensic evidence," Officer Thorpe snapped back at Sergeant Cummings. "Now don't get your knickers in a bunch. All I was trying to say is how little I know about your history after all these years. If that's a slam on anybody, it's a slam on me."

"It's not a slam on you, Robert. I should have trusted you more and opened up a little. It's just that in our line of work, when you start to get close to someone, they usually end up on the obituary page. And that hurts." Robert offered up as a reasonable response to both their cases. "Well! George starting now, I'm going to make it a point to learn something new about you each week. Until I really know who Officer Frank Thorpe is."

George turned toward Robert and smiled. "Ditto!"

As they cruise toward the donut shop, Sergeant Cummings was thinking out loud. "How could anyone walk away from a barrage like that without a single scratch?" and "Why would someone try to off such a powerful gang leader?" Sergeant Cummings continued his thoughts aloud, knowing that Officer Thorpe would jump in with a solution eventually; he always did. "Is someone trying to take over the city gangs?"

"That's a scary thought, Sergeant Cummings," Officer Thorpe volunteered.

"Yes! I know! That is a very scary thought. It would mean an awful bloodbath in the city of New York. A lot of innocent people would be killed!"

Officer Thorpe shared his frustration. "Here we are, just ten days before Christmas, and some son of a bitch has decided he wants to be king of the hood. I hope those detectives are worth the weight of their gold shields. Because I'm afraid we have some serious trouble brewing tonight!"

"That's affirmative, George. They'll come up with something—they always do. Besides, you can bet the captain's riding up their asses like a thong. You know he's going to want some answers before he goes to the chief." Sergeant Cummings offered to settle George down.

Sergeant Cummings and Officer Thorpe pulled into Darnel's Donuts and saw a line of police cars in the parking lot. They got out and walked up to the entrance and noticed Officer Buttes and Officer Rush sitting just inside with two cups of coffee and two sugar donuts waiting for them. as they entered the donut shop. Officer Buttes said, "There's our two crime fighters. Sit down and tell us all about it. We heard the chatter on the radio…so it turned out to be more that an abandoned vehicle after all."

Back at the scene of the crime, Captain Belmont ran up the steps into the mobile forensic laboratory. "Jamison, what do you have for me I need something to tell the chief."

Bill Jamison, Chief of Forensic Science for the city of New York, looked at Captain Belmont with a puzzled expression. "Well, Captain, I'll tell you straight up I have no idea how these body parts were taken from these men."

Captain Belmont was frustrated with the response he got. "What do you mean, Jamison? I need some answers quick!"

"Look at this, Captain. Tell me what you see." Jamison moved aside and let Captain Belmont look through the microscope on the examining table.

"I see several round cells meshed together like a zipper."

Stepping back from the microscope, the captain looked to Jamison for answers. "Exactly, the key word here is meshed together. Just like pure molecular separation. There's no lacerations, tearing or trauma of any kind on these tissue samples."

"Here! Maybe this will help you understand what's puzzling me. If I cut the skin of this apple with a surgical knife like this." Jamison cut the apple with a sharp knife from his table. "I put it under the same microscope at the same magnification." He places the skin of the apple under the microscope. "Now look and tell me what you see." The captain looked through the microscope. "I see the skin of an apple and the edges look as though they have been torn very neatly."

"And you are right—that's what a surgical incision looks like. A series of very minute tears or cuts. But when you look at this tissue from these men's wounds, there is no tearing at all, not even a bend in any of the tissues. They are perfectly separated at the molecular structure of the cells. They've come apart at the seam." Captain Belmont took another look through the microscope to compare the difference that Jamison was explaining to him.

"Okay then! How did these men die?" Jamison looked at the captain as though he was asking a rhetorical question. After a few moments of uncomfortable silence, Jamison responded.

"Sir! These two men died of massive blood loss due to their wounds. Their hearts, a circular portion of the chest plate, and spinal column separated completely from their bodies, and they died." Jamison, showing some frustration at having to repeat himself, held up the skin, chest plat, spinal cord, and heart from one of the victims as he explains to this to the captain.

"What instrument or weapon could have done this to these men? A shotgun maybe?" the captain continued, hoping someone will explain to him what killed these men.

"Sir, there is no weapon or procedure for that matter known to the scientific community that can duplicate what happened here. It's just not possible."

"A laser, perhaps?" the chief offered up as he entered the laboratory. Everyone stood up and greeted the Chief of Police. Chief Charles Grissom had been the police chief of the NYPD for fifteen years and was one of the most celebrated figures in the city.

"Hey, Chief, what brings you out to the crime scene at this hour?" Captain Belmont said nervously.

"Settle down, Captain. I heard some interesting chatter on the radio earlier and thought I'd come out here and see firsthand what we have. It sounds pretty bizarre."

"Well, it's extraordinarily bizarre, Chief," Jamison said as he sat back down at the examining table. Jamison updated the chief on what they have discovered up to now. "Jamison, I heard most of your explanation outside as I waited for the right moment to pop in. So do you think a laser could have done this?"

"No, a laser would cauterize the wound, and there is no evidence of that at all."

"What about a sonic weapon of some kind?" the chief continued.

"No sir, a sonic weapon would leave tears to the tissue and trauma to the affected tissues. There's no trauma at all to these wounds. They simply separated themselves from the bodies. I don't know how this is possible, but that's exactly what we're looking at."

"Well then, how about a chemical weapon of some kind?" the chief threw out for entertainment value.

Jamison thought for a moment as though the chief might have hit on something. "Well, that would be a possibility if there was a weapon that could discharge a chemical or biological agent that forces the separation of healthy molecules through ten inches of diverse density. But to be able to disperse that through the body of a man would take a great deal of force, especially to get through the bone. There lies the problem because any weapon capable of that much force would cause trauma to the tissue. That leads us right back to square one."

CHAPTER 11

Stranger in a Coat!

Meanwhile, somewhere on the other side of town, Mary Carter and her fourteen-year-old daughter, Betty, are shopping for the perfect gift for dad. They decide to get him a new leather briefcase with his initials engraved on it. "This is the perfect gift! Daddy will be so surprised," Betty assure her mom, while Mom fiddles with her purse, counting out the exact change. She pays the clerk at Harman's Leather Goods then motioned Betty to grab the door.

The chilly air and blowing snow gave Mary and Betty a shiver as they entered the parking lot. Mary looked toward the car and noticed two young men horsing around near her vehicle. Instinct tells her to take Betty back inside the store until they leave.

She hesitates for a moment in hopes that the two boys will move on. Betty grabbed Mom's arm and pulled it. "Let's go! It's cold! We need to get home and wrap it before Daddy gets there." Betty encouraged her mother toward the car with one more plea. "Come on, Mom!"

Mary reluctantly yielded to her daughter's prompting and headed for their vehicle with Betty on one arm and the briefcase in the other. She shook off some of her concern when she noticed a couple making out in a car two spaces from theirs. As she approached her vehicle, she also noticed an older gentleman loading groceries into his truck, just a few spaces behind her car.

Mary fumbled in her purse for her car keys as she arrived at the driver's side door. She managed to extract her keys and unlocked the doors, just as Betty walked around the passenger side of the car. Mary sensed they are not alone. Suddenly, the two young men she'd seen before appeared and grabbed them. Mary noticed that one of them had a knife and screamed, "Help! Someone please, help us!"

The older gentleman who was previously loading groceries into his truck quickly jumped into his vehicle and sped off. The amorous young couple started their engine and drove off. Mary screamed again, looking directly at the couple as they drove away. Mary admonished them as they passed, "How could you do this? Help us, please!" Betty's plea for help landed on deaf and uncaring ears.

Hopelessly she screamed out again, but her cries seemed to be absorbed by the wall of falling snow. Jeffery Spencer, the smaller of the two hoodlums, puts a crowbar across Betty's throat and holds her down firmly while Chuck Richmond, the older, bigger boy, continued to struggle to get control of Mary.

Mary pleaded with the two, "Please! Don't hurt us, you can have anything you want. Just…please don't hurt my baby!" Chucky, still struggling with Mary, hit her twice in the face with the butt of his knife.

"Shut the fuck up, bitch! Or I'll have to cut you!"

With each blow, Mary lost more of her will and energy to fight. Finally, she yielded control to her assailant. Chucky impatiently demanded all her money while holding the knife to her throat.

"Give it up now! Bitch!" Mary gave him $329, what was left of her Christmas shopping money.

Chucky demanded her jewelry. As Mary complied, she asked, "What are you going to do to us?"

Chucky smiles grimly and replied, "That depends on what else you have to offer, honey!"

He rubs his free hand up and down her neck and over her breasts. Betty screamed, "Leave her alone! Help! Please, someone…" Betty gurgled as Jeff pushed the crowbar tight against her neck and opened the back door of the car.

He pushed Betty into the back seat and forced her to lie down. Betty struggled with him for a moment until Jeff hit her in the head with the crowbar.

Jeff yelled to Chucky, "Come on, Chucky, either fuck her or shut that bitch up! She's killing the mood, man!" He rubbed Betty's bleeding face with the back of his hand and smiled. Then he ripped her blouse apart and loosened her bra.

Betty struggled against his lewd advances, and he hit her again with his crowbar. She felt dizzy as blood rolled down her face and lost the strength to resist him. He massaged her breasts as she bit into her lower lip, stifling her instinct to scream again.

Mary panicked and summoned the strength from deep within her. Mothers had an energy reserve that lay dormant until needed to protect their young. Courageously she drove her heel into the top of her assailant's foot. Chucky screamed in pain and collapses to the ground, loosening his hold on Mary. She then kicked Chucky in the balls with all her might.

Chuckie's face went pale as he gasped for air and then let out a loud scream, filled with all the profanity he can muster. He placed his hands on his crotch and curled up into a tight ball…whimpering.

She reached for the handle on the back door of the car. Jeff reached out and locked the car door from inside. Mary desperately attempted to open the door with both hands on the handle, pulling with all her strength. She quickly realized her efforts were in vain; she slapped and pounded the rear window, pleading to Jeff, "Please, for the love of God! She's only fourteen years old! Please, don't do this! Let her go!"

Jeff looked up slowly into Mary's begging eyes as though he were giving some thought to her plea. Then he smiled and flipped her the finger and rode her daughter like a bucking bronco. He licked his middle finger and rubbed against Becky's exposed nipples, sliding it slowly to her belly button and down into the top of her jeans.

Mary lost control; she grabbed her shoe and started beating on the glass with the heel of her shoe. "You sick, pathetic pig, I'll kill you!" Jeff continued to smile at her as if to dare her to enter the car.

Chucky got to his feet, his groin still aching; he grabbed Mary's feet and pulled them out from underneath her. She smashed her face on the lip of the car door as she fell helplessly to the ground. Her head hit the snow-covered concrete as her body reached the ground. She lay unconscious face down in the snow.

He pulled his belt off and fastened it around Mary's arms so they were outstretched over her head. He tightened it around her elbows and head, making them useless to her. Rolling her over, he ripped her blouse open and cut through her bra, exposing her breasts. Feeling victorious and raging with anger, he fondled her breasts forcefully. "You're going to pay for that, bitch! How's that? You like that?

"Oh, baby! I'm going to give you something you will never forget tonight! You never got this from your old man!"

He rolled her over onto her face and lifted her skirt and pulled her panties down to her ankles. He quickly undid his pants and pulled them down with his briefs. He looked in the car and gave Jeff the thumbs-up, then he slapped her bottom firmly several times.

Mary squirmed back and forth in the snow, moaning. Chucky laughed at her and rubbed her butt cheeks as he moved closer for contact and penetration. Mary brought her feet up sharply, violently striking him in the middle of his back and knocking the wind out of his sails. He flew forward over her head.

Chucky gets up slowly, catching his breath. "You've taken all the fun out of this for me, bitch! Now I'm going to take something from you!"

He put his foot in the middle of her back, holding her down while she flopped helplessly like a fish on the dock. Then he shoved the blade of his knife into her side, slowly forcing it between her ribs and deep into her lung.

She screamed in pain and started to gurgle, drowning in her own blood. Her feet fell slowly to the ground. Chucky sat back down on top of her, tucking his penis and balls between the cheeks of her ass to keep them warm. Then he leaned close to whisper into her ear, "I want you to know that when I'm done with you here, I'm going to do the same thing to your daughter." He strained to push his knife into her other side the same way. She jerked several times and then went limp.

"You stupid bitch, I knew I was going to kill you and fuck you the minute you walked out of the store." Her body started to quiver with muscle spasms, and Chucky got aroused.

Betty heard her mother gurgling and then go silent. She cried out, "Mom! Oh no! Please stop, don't hurt my mom. I'll do anything you want. Please leave her alone!" She struggled to see if her mother was okay but couldn't move under the weight of her attacker.

"Don't worry, little girl. We won't hurt your mommy anymore… and in a few minutes, you can be with her…okay?" Jeff pulled his pants down and pulled Betty's jeans off. She lay there quietly, too scared to move or scream anymore. She lost her will to resist the assailant and fell into a conscious stupor, tearfully watching as Jeff finishes ripping her clothes off.

Jeff got ready to penetrate Betty and announced to her, "Your mother would be proud of you today, honey! Because today I'm going to make a woman out of you!"

Suddenly, Chuck disappeared in a flash, and Jeff lost his balance in the back seat as the car was jolted violently with the sound of a tremendous thud. Jeff yelled angrily, thinking that Chuck was the culprit. "Will you knock it off, asshole? I'm not through…" Jeff's voice winced in pain and rose several octaves as he felt a vicelike grip on his genitals.

Jeff disappeared out the side door of the vehicle in the blink of an eye. He flew out backward and was smashed into the car next to them. His body came to an instant stop with a loud popping sound. The silent winter air was filled with another loud thud, and then he splashed into the sludge on the ground.

Betty, barely conscious, noticed the figure of a tall stranger standing outside the car. "Come forth, young Betty. All is well now!" A calm, soothing voice echoed through the car. Betty sat up, grabbing her garments to cover her naked, shivering body. Betty's fear left her as she leaned next to the door, looking out. She saw the image of a tall stranger in a long overcoat with a wide-brimmed hat on his head.

Betty stepped out of the car while the tall figure walked over to the other side of the car, looking down at the ground. She followed him and saw the lifeless body of her mother, bound and gagged, laying naked in the snow.

"Oh my god! Is she dead?" Betty blurted out and started to run over to her mother.

"Fear not, young lass! All will be right in a moment. Trust in the Lord."

Betty was unable to move forward toward her mother and was calmed by an unseen force.

The two hooligans were dangling in the air with their pants still around their knees, unable to speak or move. Their eyes were wide open, and their minds were conscious. They just couldn't move at all like little rag dolls dangling in the air, hovering several feet over the scene.

Betty was scared at first, thinking it was a trick and fearful that they would pounce down on her any moment. The tall figure sensed her fear and comforted her. "They can't hurt you any longer, dear!" Then he tapped his two index fingers together in front of him, and the limp, rag doll images of her assailants banged together like clacker balls at the end of a long string.

Betty laughed at the image and smiled at the tall, dark stranger. "What's your name, mister?" she asks politely.

"John," he replied, returning her smile.

"They can't hurt me anymore?" she asked inquisitively.

"No, honey, they can't hurt anyone anymore. They still have all their senses so they can witness and feel the pain they have caused you and your mother. But they are quite helpless right now."

Betty looked at the two boys hanging in the air with pleading faces. They both returned her glance with pitiful and shamed looks. She shot them the finger and shouted, "Fuck you, assholes!" John quickly looked around to see if anyone else noticed her comments.

The ground started to shake violently, and two larger-than-life fingers protruded from the ground, jetting upward. Middle fingers of an enormous hand, each about seven inches in diameter, traveled swiftly into the buttholes of the two villains dangling helplessly in the air.

Unable to move a single muscle, their eyes rolled back and forth in their sockets, and their muted voices squealed like a tortured pig.

John turned his head, trying to withhold his laughter. After composing himself, he turned back and snapped his fingers. You heard a couple of gentle pops as the two giant fingers withdrew and disappeared. "Betty, you must be careful with what you say right now. Please let me handle this."

Betty looked at John and agreed. "Okay, but it's going to be hard for me to keep quiet!" She glanced back at Chucky and Jeff, seeing their eyelids fluttering and tears running down their cheeks.

Not yet satisfied that they had gotten what they deserved, she asks John, loud enough for both the hoodlums to clearly hear. "So! John? You mean that anything I say will be created literally?" she asked with childlike innocence.

"Oh yes, dear! It's created very literally, and their senses are much more tuned right now, so they feel it three times more intensely. So please don't say anything, and just let me handle this. Everything will be fine. I promise you," John explained sincerely.

Without hesitation, Betty responded, "You mean if I were to say I wanted them to feel a red-hot iron rod…" John put his hand gently over her mouth to suppress her from completing the sentence. At the same time, Chucky and Jeff's eyes got enormous while listening to Betty start her sentence. Then they start to relax, noticing John had covered her mouth before she could complete the sentence.

Unfortunately for the two men, enough of the sentence was uttered to complete a thought. Their eyes start to pop out of their eye sockets, looking down at their penises. The men's rods had risen enormously to meet their gaze. Exaggerated in size, both their dicks appeared as two huge, red glow sticks, filled with fire and brimstone. Water poured out of their tear ducts, and they whimpered like puppies.

"I'm sorry, young Betty, but I have work to do. You're going to have to remain silent for a while." John pulled his hand away from her mouth, and she remained silent. He looked at the two men with their two fiery towers of molten lava and snapped his fingers again. Instantly the organs fell limp while charred flesh dripped to the ground.

John looked back at Betty and said sternly, "You realize the damage caused is irreversible—they can no longer procreate!" Betty looked up and smiled silently; comforted in that knowledge, she felt happy. Then sadly, she glanced down at her mother, wishing she were here.

John walked over to Mary's body and kneeled next to her. He put his hands on her head and recited a prayer. As he completed the prayer, he stood and walked back to Betty's side. He put his arm over her shoulder and called out to Mary, "Behold the power of the Lord! Come forth, Mary Carter, and see your daughter!"

Mary began to breathe and sat up in the snow. The clothes that were torn, cut, and removed from her body were back in one piece. Even her hair was in order, and her wounds are healed. She stood up and looked at the man and thanked him. "Thank you, mister!"

"Don't thank me, Mary Carter. It's your faith that has made you whole. Thank the one who sent me." John glanced upward. Mary followed his cue. "Thank you, sweet Lord, for the safety of my daughter and for sending this kind man to help us."

Chucky and Jeff continued to witness the events unfolding before them and were ashamed. They hung their heads in despair, and tears of joy flowed as they witness Mary embracing her daughter. "Betty, are you okay, honey?" Mary asked as she squeezed her daughter close.

"Yes, Mother. I'm fine now!" They hugged each other and kissed.

Betty walked over to John and hugged him. "Thank you, John! You're a nice man." John returned the hug. "You don't have to thank me, Betty. My thanks is seeing you and your mother back together again in time to share a wonderful event."

"Can you teach me how to do that?" Betty asked, referring to his ability to help them.

"I can't but there's someone who can, and he rests within your pure heart, Betty. When you seek him with a sincere and pure love, he will bless you with many gifts—treasures the eye can't see, the tongue can't speak, and the ear can't hear. Such is the way of the Lord God Almighty. Amen!"

Mary joined her daughter and John and gave him a hug too. He whispered into her ear quietly, "Your husband will love the briefcase you purchased. He's always wanted that one."

"It's time for you and Betty to go home. You need to wrap that gift and prepare for your husband's arrival. Of this incident you and your daughter will remember nothing as though it never happened. Now! Go in peace and may the Lord keep you well…always!"

With that said, John snapped his fingers, and both Mary and Betty arrived at their car with the briefcase in hand. They entered the car and drove off. They were excited to get home and giggling as the car rolled into the street. John watched as they drove away. Betty turned, smiled at John, and winked at him.

John smiled and turned back to the two whimpering boys. Their torched penises and scrotums had fallen off and lay smoldering in the snow below them. Their assholes were the size of moon craters. He looked at them, shaking his head. He snapped his fingers, and the two boys floated to the ground, still unable to move.

Looking at them, he said, "What you have witnessed here today will be a testimony for your faith. You both have a chance to turn your lives around, but it won't be an easy journey. The parts you are missing can't be restored. You will have to live the rest of your lives knowing that you brought this vengeance on yourselves."

He released their silence to hear what they had to say. Chucky was first to respond. "Mister, I'm sorry for what I did to that woman. I've been a creep most of my life. It's all I know how to do. But if you let me go, I'm going straight over to St. Paul's Cathedral and confess my sins and throw myself on the Lord's mercy."

John waved his hand, and Chucky was reunited with his flesh body, and he instructed Chucky, "Be off, and for the sake of your soul…honor thy promise."

Chuck agreed, "I will, I promise you I will. And again, I'm so sorry for the pain I caused that woman and her daughter." Chucky put his pants back on and ran toward the cathedral.

John looked at Jeff, and he too expressed his sorrow. "Mister, I'm only sorry that I wasn't able to fuck and gut that bitch before you showed up. Old man, there isn't anything you or anyone else can do to get me to run off to some damn church and beg for mercy. Chucky is a fucking-pussy."

Looking at the boy, he shook his head sadly. "I'm sorry to hear you say that, son!"

Jeff reached down and pulled up his pants, slowly trying to hold in the screams of pain. "Fuck you, asshole! I ain't your son!"

John turned and walked into the night, disappearing in a wall of snow. Jeffs got his pants up, just barely, and started to walk away very slowly. He glanced down and noticed how badly he was bleeding. "Shit! Fuck you, bitch! When I get my hands on you, I'm going to cut your tits off and use them for a fucking Band-Aid!" He shouted into the cold night air.

Each step he took through the parking lot, he finds himself losing energy, getting weaker and weaker. Then he collapsed onto the snow-covered parking lot, looking back on a trail of blood in the snow. Slowly he lost consciousness as life escaped his body.

Moments later, he found himself standing over his own body, frozen in a pool of blood. "You useless piece of shit! What good are you?" he shouted out as he kicked his own body several times.

Jeff heard screams and wailing coming from all over the area. "Kiss my ass!" He started to walk away from his own body. A large figure with a cloak and hood covering his shape started toward him. The figure was eight feet tall and covered from head to toe with a full-length black hooded robe.

Jeff tried to ignore the figure and pointed to his own body in the snow. "He's over there, asshole. Keep walking that way, and you'll trip over him." Jeff stood to one side to let the hooded figure pass.

The figure looked at the body in the snow and stopped next to Jeff and shook his head. Then it laughed like a hyena and threw his robe to the ground. "I prefer you! Asshole!"

The hideous creature stood next to him with a large mouth filled with grizzly teeth protruding from his stomach. His head was one large, bloody scab. Jeff fell back at the grotesque sight. "No, tough guy, don't quit now. I was enjoying the show." The scab-headed creature turned to look at Jeff's body resting in a pool of blood steaming in the snow.

The belly of his frozen body started to swell up like a pregnant woman. Then it burst apart, and thousands of scorpions crawled out and started to sting the body and rip it apart as though he were still alive.

Each sting was more agonizing that the one before, and with each rip, Jeff watched another piece of his body disappear in front of this huge and hideous beast. The scorpions brought his body over to the monster piece by piece and tossed it into its mouth as he slowly and painfully devoured him.

Jeffs spirit stood helplessly while his body was slowly consumed, slowly shrinking him down to nothing, leaving no piece behind, including his charred penis. Once all the pieces were dumped into the mouth of the beast, his body became translucent, and Jeff's body was reassembled in the belly of the creature. His spirit was sucked into the belly of the demon and reunited with his body. The pain was intense and caused Jeff to scream and gnash his teeth. With each passing moment, the pain intensified.

The scorpions continued delivering excruciating pain with every sting. He screamed in agony, floating in the monster's bile and brimstone. The monster burped, pulled up his robe, and sauntered out of the parking lot with Jeff kicking and screaming.

Chapter 12

Fast Cash!

It was a beautiful, clear October sunny morning in Phoenix, Arizona. The sun was bright, the sky was blue, and the city was one hour on a normal busy day. The Phoenix Civic Center was buzzing with several different events. A boat show and fishing exhibition was setting up for the weekend at one end of the center.

A home arts and crafts show was preparing to go live on Channel 3's Good Morning Arizona program. Out in the front, near the pavilion, live performers were setting up to show off their talent in hopes of getting discovered.

In room C of the Phoenix Civic Center, people were milling around a cloth-covered sign-in table with three lovelies, well-dressed, professional-looking women named Tiffany, Angela, and Sheila. They were sitting on the opposite side of the table busy filling out nametags for the guests. There was a list for people to sign as they entered the large open room where ten tables were set up in the shape of a large U. The function seemed unusually small, given the capacity of the room was 150. There were only about forty or fifty people inside.

Harvey and Mildred Button, an elderly couple from Sun City, dressed in their finest polyester outfits, slowly approached the entry door, carefully double-checking the address in the newspaper ad they were carrying. Mildred priggishly nagged her husband, "Harvey, I'm not saying it again—this is the place. Just come on in and be quiet."

Harvey mumbled loud enough for Mildred to hear him, but being careful not to talk over her, he says, "Are you sure this is the place, honey? I thought there would be a lot more people here."

Behind them was a young man dressed in wide-leg jeans that seemed to be belted around his thighs with his boxer shorts pulled up around his waist. The man had sunburst-orange hair with a silver stud in his chin, carrying a brightly painted skateboard and sporting a backpack. He walked as though he was in no hurry to get anywhere. His name was Spinner, a nickname given to him by his friends because of how fast he could spin on his skateboard. His legal name, shown on his driver's license, was Bryan Smith. He moved to Arizona from Dover, Delaware, to attend Arizona State University and was in his freshman year. His major was girls and parties. He was hoping that skateboarding would become an Olympic event one day.

Inside, several people chatted about everything but the weather or why they were there. Some stood next to the coffee and juice table against the wall, engaging in a conversation about fresh cinnamon rolls. Others mingled around the tables, trying to find the best seat in the house, claiming their spots with notepads and personal belongings.

The door was closed at 9:00 a.m. A handsome, nicely tanned fellow with a muscular build walked over to the over to the open end of the tables and invited everyone to take a seat and make themselves comfortable. He was forty-five years old, dressed in a stylish blue suit with shiny leather shoes. His brown hair was neatly combed back and showed a distinguished hint of gray on either side of his head. His body language presented him as a man who was accomplished, confident, and successful. He was well-spoken and comfortable in front of large groups of people.

Everyone moved quickly to his or her seats to sit down and get comfortable. "My name is Mr. Palmer Gilmore. I'm president of Bonded Drivers of America or BDA to keep it simple. First, I want to thank you all for responding to our ad in the newspapers. BDA, as we are known in our business, is an agency that hires non-professional drivers like you as needed by our clients to move vehicles from one place to another."

Behind the speaker, a projector lit up the Bonded Drivers of America banner.

"You can call me Palmer. After all, my mother called me that for thirty-eight years. She also called me by other names that were similar. For instance, I recall once she called me home from the park where I had been playing with my friends. She noticed my pants were hanging down on my butt about as far as this gentleman's pants are." He pointed to Spinner, who managed to wrangle a seat up at the front of the table. He smiled and patted Spinner on the shoulder.

"Unfortunately, I wasn't too keen with the new styles back then and was displaying what is now comically referred to as a plumber's crack." Everyone broke into light laughter. Mr. Gilmore finished his story. "My mom was quick to announce to all within earshot, Plumber Palmer! You get back to the house this instant and pull your pants up where they belong. After college, you can decide which profession you wish to pursue but not on the playground, young man." More laughter erupted.

"As you can plainly see, I chose a career unrelated to physical labor." He got another chuckle from everyone and continued. "We run ads periodically as a need presents itself through our clients. The client in this case has chosen to remain anonymous. We get paid very well by our clients to honor their wishes. So let's get right to the heart of why we are here today."

Everyone picked up pencils and pulled pads of paper in front of them that were supplied by the company. Palmer started off by telling them, "This is perfectly legal and above board. No one here is committed to stay. If, for any reason, you're not comfortable about the money or the job, you may leave at any time, no questions asked." Then he paused and looked around the table to see if anyone was having second thoughts before he continued.

Palmer explained, "Most of you in this room will be bussed to a small town in Southern Mexico, where you will be handed the keys and papers to a car or truck. You will be fed—very well I might add—and roomed for one night. Then you will each drive your vehicles back to the US border town of Nogales, Arizona. You will drive to MO's garage and drop off your vehicle. You will get on a bus that will be waiting for you. The bus will bring you back to Phoenix and drop you off here at the Civic Center. Then come back to this room, and you will be given an envelope with $1,500 cash in it. It's that simple."

Mr. Gilmore nodded to the projectionist to put up the next slide. He waited a moment for people to absorb the information just shared. Then he continued, "The cars and trucks you will be driving are repossessed vehicles, recovered from people that borrowed money to pay for these vehicles and then defaulted on the loans that were written by our clients."

Harvey raised his hand and asked, "Why not truck them into the United States? Why do you need us?"

"That's a good question, Mr. Button." He read his name of the name tag provided earlier. "We don't have a trucking agreement with Mexico and the United States to transport any reclaimed goods. So the only legal way to move repossessed cars and trucks between our two countries is the old-fashioned way. The company titles and registers the vehicles you will be driving in your name. The papers in your glove box will reflect that."

There was a little mumbling going on at the middle of the table and finally another question. "What exactly do we do if we are stopped at the border?"

"That's another good question, and I'll take a moment to answer it for you. But before I do, I want all of you to hold up the information package that Miss Pepperton passed out earlier. Does everyone have it?" He paused for them to hold it up.

"Good. In that package is the answer to all your questions. We have done this a hundred times before, and the questions you have presented and the ones that you want to present are all answered in that package."

"Now back to the question. If you are stopped at the border or anywhere along the trip, just provide the papers located in your glove box to the US or Mexican officials. Those papers include the title, registration, and even insurance card showing you as the legal owner of the vehicle. And you are, at least until you get to MO's garage. That's where your ownership of the vehicle ends and the company reclaims its ownership."

Palmer could see that people weren't getting it. "Folks, please listen carefully for a moment and I'll try to break it down a little more for you. The company that lent the money for the purchase of these vehicles owns them, according to the laws of both countries. The people that borrowed the money from the company to buy the cars and trucks have defaulted on those loans. In Mexico, the consumer no longer has any legal right to the vehicle. The company owns it until the financial obligation is met and may reassign ownership of that vehicle when the loan is in default. Basically, the consumer has no rights to the vehicle if they haven't paid the loan off." Palmer took a moment to ensure everyone was on the same page.

"The company can and has legally assigned ownership to you. So once you have the keys to the vehicle, it's legally your vehicle. When you cross the border into the United States, you are still the legal owner. But you are agreeing to turn ownership over to the company in exchange for $1,500 by signing your name on the back of the new title. That will take place at MO's garage in Nogales. Now does it make more sense?" Everyone nodded in understanding.

To cut the questions short, Palmer interrupted and said, "Listen, please. If anyone is uncomfortable with driving one of these repossessed vehicles out of Mexico and into the United States, you may leave now. That is what this briefing is all about, to make those that choose to drive for us are comfortable with the trip.

"Please take a few minutes to discuss what you want to do, and we will resume shortly."

Everyone turned to each other at the table and started discussing what they had heard. A man in his thirties named James Evans looked around the table quietly. He spotted a young, blond-haired woman two chairs down from him. She looked as though she was alone and was sitting quietly by herself. James pushed his chair back and walked over to where she was.

"Hello! My name is James Evans, so what do you think of all this?"

She looked up at him, slowly checking him out. Then she said, "Well, I don't really know what to make of it. I'm just like everyone else here." She looked around the table. "I need the money. It sounds good, so here I am." When she made eye contact with James, she introduced herself. "My name is Sherri Cline." She offered him a seat next to her, which he gladly accepted.

He looked around to see if anyone was listening to them and then leaned over close to her. "Honestly, I think there is more to it. I mean please, $1,500 for two days work! I'd be very surprised if there wasn't a bomb in every vehicle. I do need the money, though—I'm trying to get a few bucks together to throw to the bill collectors. I've been out of work—at least steady work—for several months now and my savings are depleted. This is the quickest money in town. If it does turn out to be legit, I'm going to be a permanent driver for them."

Sherri continued to listen while staring into his beautiful eyes. After he was done talking. James broke her gaze and asked, "What?"

She shook her gaze off and responded, "Yah! You're right. It does seem to be unusually easy money, but that's why I'm here…easy money. At least this job doesn't require me taking off my clothes. I've had my fill of those types of jobs, and they suck. I'm going to use my money to finish paying for my nurse training and hopefully get a job at Desert Samaritan over in Mesa."

"Really! That's great!" James replied to keep the conversation going. He was taken by her beauty. Sherri was 5'11" and 140 pounds. She had sky-blue eyes and cute dimples when she smiled. All her parts were in the right place, and she was well-proportioned.

Mr. Gilmore announced "Time!" as he walked back over to the tables and looked around at those remaining. "Well, I see we didn't lose too many—I count thirty-two heads still here. Is that what you have, Miss Pepperton?"

"Yes, that's what I have—thirty-two drivers, Mr. Gilmore." Then she turned and walked to the back of the room. Palmer smiled and walked slowly around the tables, checking everyone out.

As he walked around the table, he handed yellow cards to Harvey and Mildred Button. Then walked over to another man named Mike Crow who was dressed in combat fatigues and had long, unwashed brown hair.

Finally, he gave the last card to James Washington, a black man who had arrived with some friends, Thomas Milky and Eugene Farmer, who were sitting next to him. Mr. Washington had been joking about the whole process and insisting this was a drug run for the Cambodians. He also had asked for more money thinking he might get it because of the drugs.

After handing out all the yellow cards, he explained, "Those of you who have received a yellow card are free to go. We have too many drivers and we have randomly selected you to drive next time. I'm sorry you can't go on this trip, but we will keep your name and number on file for future drives. Thank you again for coming out."

Then he paused to allow them time to be escorted out of the room by Tiffany, Angela, and Sheila, the three ladies that were at the sign-in table.

After they left, he continued, "The rest of you can go get something to eat and be back here at 2:00 p.m. There will be a bus waiting for you in the north parking lot. We have planned, for those of you who drove your own vehicles, to have them held in the northern corner of the parking lot. There will be no charge for this, but you will need to leave your keys with the attendant in case they have to move your vehicle. Thank you all for coming, and we will see you back here at 2:00 p.m. sharp."

Mr. Palmer Gilmore and his entourage headed for the door, while Tiffany, Angela, and Sheila started to clean the place up.

James asked Sherri where she was going for lunch. Sherri responded, "I'm not sure. Where are you going?" James put his hand to his chin and tapped it with his finger for a moment. "I know this great hotdog place on McQueen and Broadway about twenty minutes from here. They're the best in town. It's called Ted's. Have you ever eaten there?"

Sherri tapped her chin with her finger, mimicking James, and then said, "No! Never, let's go!" They grabbed their stuff and headed out the door.

Outside, Palmer jumped into a waiting limousine while dialing on his cell phone. "Hello! This is Palmer. Is Manuel there?"

"Hey, Manuel, I have the drivers. Yes! Twenty-eight…just as you ordered. These guys are just as stupid as the last bunch—we checked them out. Only a few of them had ties to the Valley. We let them go with a call-back card. Yes! They will be here at 2:00 p.m. sharp and should arrive at the hotel tomorrow around 5:00 p.m. Yes! I have the perfect guinea pig for Dr. Swanson."

"He's a nineteen-year-old kid that fits the bill to a tee. I'm going to get him now, and I'll run him by Doc's and pick him up just before the 2:00 p.m. bus ride. Okay! Okay! I know he'll be there."

"Listen, you promised a bonus on this one, and I want to make sure you understand I'm done after this one. It's getting way too hot around here. Hello…hello…you son of a bitch!" The phone goes dead.

Palmer yelled at the driver to go by the parking lot and pick up the orange-haired kid. The driver took off and spun around the corner to see Spinner playing with his skateboard on the walkway.

The car pulled up to the curb, and the back door swung open as the car came to a screeching halt.

CHAPTER 13

Visit to the Doctor

"Hey, Spinner!" Palmer yelled as he stepped out of the car. "You got a minute? I need to talk to you." Spinner jumped off his board, a few feet from the car, and shrugged. "Yeah! What's up?"

Palmer looked down at his shoes and wiped them on the back of his pant legs. "I kind of felt bad about picking on you in there. Why don't you let me spring for lunch?"

Spinner lit up with a smile and said, "Shit, yeah, you got a deal, man!" and jumped into the car with Palmer. The car pulls off down Third Avenue toward the Phoenix business district.

Palmer engaged Spinner in idle chitchat along the way to confirm he had no relatives or friends in the area that might be worried if he was gone for a while. Spinner told Palmer he didn't know anyone in the area, and the only time he talked to his parents was when he needed money.

Palmer smiled and said, "Well, here we are. I hope you like this place. It's not much to look at from the outside, but it has great food, and the service is to die for."

They pulled up to an elegant strip club called the Cherry Blossom. "Spinner, why don't you go in and pick out a table for us? I'll be there in just a minute. I must make a quick call. Tell them Old Man Palmer brought you."

Spinner jumped out of the car, leaving his skateboard and backpack inside, and headed to the front door. As he got close, he noticed a sign on the glass that said, "Adult entertainment, must be 21, topless waitress, free lap dance for first-timers." He turned back toward the car where Palmer was talking to someone on the cell phone and screamed, "Awesome, dude! You're the best!" and flashed him two thumbs up and a huge smile.

As he turned and entered the club, two large men who looked as though they spent every minute at the gym working out with five-hundred-pound barbells greeted him. He said to them, "The guy in the car told me to tell you I was with Old Man Palmer."

Both looked at each other, uncrossed their arms, and shook his hand. "Hi! I'm Eric, and this is my partner, Butch. Welcome to the famous Cherry Blossom Club for men. Is this your first time here?" Eric asked politely.

"Yeah!" he replied with a huge smile, bobbing his head up and down.

"Well, sir, you are in for the time of your life. Come this way please."

They led him through two red leather doors, trimmed with golden studs, with a half-moon window on each side in the shape of a breast. As they entered, Spinner took a big sniff and asked, "What's that smell?"

Butch responded, "That's the scent of fresh cherry blossoms in bloom. We have it sprayed into the return ducks every twenty minutes. Do you like it?" He nodded in approval.

Spinner stopped just inside the door and looked around the club. It took a minute for his eyes to adjust to the change in lighting. Then he scanned the club's layout; it was quite impressive.

To the left of the entrance was a wall-length bar with under-the-counter lighting, giving off a blue glow. It offered just enough light for the bartenders to mix drinks without distracting from the ambience of the room.

The ceiling was dimly lit with reflective lights pointing upward toward large, white half-dome-shaped circles, recessed into a black background. There was glitter throughout the flat, black ceiling to give the impression of stars.

The club was designed as a trilevel circle that sloped to the center of the room, where a large stage commanded the attention of the room. The stage looked like a clover with a golden pole at the end of each leaf.

Mirrors above the stage provided the most hidden booth a clear view of the dancers. On each of the three levels, there were semi-private and open, chest-high booths surrounded by live plants mixed with cherry blossom branches. It was an appealing design.

As the continued to walk slowly toward the stage, he started to make out the girls dancing with nothing but a G-string. "Holy shit! You have got to be kidding me. I've died and gone to booby heaven. This is so cool!" Spinner said to his two burly escorts without taking his eyes of the stage.

"We're pleased you like it, sir. My partner and I have worked hard to put this together." He finally looked at the two Goliaths and said, "Oh yeah!" while nodding in approval.

"Sir, may I get you something to drink while you wait for Mr. Gilmore?" Eric said as he seated Spinner at a table right next to the stage. Behind him, Butch motioned to a cute, large-breasted dancer to come over to entertain him.

Spinner pulled Eric closer and explained, "Hey, man, I don't want you guys to get in any trouble, but I'm only nineteen years old."

Eric leaned in closer and whispered, "Sir, we don't ask or tell here. You're a friend of Old Man Palmer, and that's good enough for us." Then he straightened up and asked again, "May I have your order, sir?"

"Yes, I'll have a Heineken Eric, and from now on, would you guys please call me Spinner? The 'sir' thing is freaking me out."

"Of course, Mr. Spinner. It would be our pleasure." Eric and Butch headed off to the bar to get him a beer and to let Palmer know he had settled in fine. Cherry Bomb, a cute stripper, came over to the end of the stage where Spinner was sitting and started to shake, shimmy, and roll. Spinner's eyes locked onto her like a heat-seeking missile as he watched every undulating move she made.

Eric grabbed the phone as they got to the bar and told Palmer, "The guinea pig is on the wheel. We'll introduce him to the good doctor shortly." And he hung up the phone then turned toward Butch and nods.

Butch slipped a white powdery substance into a frosted beer mug and filled it with Heineken and calls for a waitress. "Charlene, you see that kid over there?" Pointing out Spinner, she nodded. "His name is Spinner. He's a friend of Old Man Palmer. I want you to take this beer over to his table and present it with a cherry on top." Butch winked and smiled at her while running his finger gently under her chin.

"You bet, honey. I'll take good care of him." Charlene took the beer over to his table and put it down right in front of him.

"Hi there. I'm Charlene. This one is on the house, compliments of the boss. If you need anything… I mean anything!" She cradled her bulbous breast in her hands. "You just whistle!"

She leaned closer to him. "You do know how to whistle, don't you?" She ran her fingers through his hair, leaning far enough over so her right nipple dipped into his cold beer. "You just put your lips together and blow!" She pulled her breast out of the beer and held it up in front of his mouth. "How thirsty are you, Mr. Spinner?"

Shaking like a schoolboy on his first date, he leaned forward and licked the beer foam off her nipple and fell back into his seat, sweating. "Thank you, Charlene. That's the best damn beer I've ever had."

Charlene pulled back with a satisfied expression on her face and replied, "There's more where that came from, cutie. Now remember all you have to do is whistle." She turned and headed back to the bar, where she saw Butch and Eric high-five each other. Making eye contact with Butch, she blew him a kiss and winked.

Spinner started chugging away on his beer, figuring if he got a quick buzz on, he won't be so nervous. Soon, he'd be whistling like a canary!

The powder Butch put in the beer took effect immediately. Spinner's head fell forward and smacked the table.

Eric and Butch jumped over the counter and headed to the table to collect their prize. Eric picked the boy up and heaved him over Butch's shoulder, remarking to the few patrons in the club.

"First-timer folks, nothing to see here—next round's on us!" Then they carried the 140-pound body out of the club's double doors and into the office door opposite the club entrance.

Eric entered the office and walked over to a large dark oak bookcase behind the desk and pulled on a book titled *'The Dark Figure by Kim Richard Smith'* sitting in the center of the bookshelf.

The bookshelf slid to the left, exposing another door. Eric opened the door, and Butch walked through to a staircase leading down to the basement under the club's stage.

He put the limp body into a wooden chair in the middle of the nearly bare room. Then Eric helped him fasten Spinner's arms and legs tightly with the metal straps attached to the chair. Butch held Spinner's head up while Eric fastened a metal ribbed helmet to it. After he was secured in the chair, Eric yelled out.

"He's all yours, Doc. We're out of here." And they headed back up the stairs to the office.

"Thank you, gentlemen!" a voice echoed from behind a black curtain against the wall. Dr. Archibald Swanson stepped out from behind the curtain dressed in a white lab coat, green and yellow plaid pants, and two-tone golf shoes. Tapping a syringe full of clear liquid, he stared at the young man in the chair. "Well…well! What do we have here, another project of true persuasion? How nice for me and oh so bad for you!"

Dr. Swanson is known as a master of mind manipulation with his circle of influence. He has been called the Butcher, the Barbarian, and his favorite of all, Dr. Pain. He graduated from Boston's famous College of Medicine after six years of study and then traveled to Europe to study in the Heinrichs Scientific Research Center. There he studied under the finest doctors in the world. Securing his PhD for his thesis on "Human Intolerance to Pain."

Assuming he would be received into the medical science community with open arms when he returned to the United States. Dr. Swanson prepared a full brief and documentation on his experiments to present at the Boston Hall of Medical Science.

But on the contrary, they wanted nothing to do with him. Dr. Swanson had performed his experiments on live, mentally challenged human subjects, and that was his downfall. No one in the United States Medical Science community wanted anything to do with his study, and he was black balled as a renegade madman.

So Dr. Swanson did what any mad scientist would do. He sold out to the highest bidder and has been practicing on human beings presented by his new clients and making millions.

Dr. Swanson was totally committed to proving his theory beyond all doubt to the medical science community that shunned him. His theory states that man can easily be programmed to perform any act or function with no regard for moral judgement or physical limitations.

Pursuant to this means, he developed a small electronic receiver that could be inserted into the brain over the pain center and was triggered by remote control. Once triggered, it caused unbearable pain in the subject.

Furthermore, he developed a simple method of sensory induced hypnosis, which was easily administered without the consent of the subject. This allowed complete stealth in any given mission.

His most hideous work recently was the uni-bomber, Marine barracks bombing in Yemen, bombing of the federal building in Oklahoma City, genocide in South Africa, and the bombing of buildings in New York, Chicago, and Los Angeles just to name a few.

The CIA, FBI, NSA, MI6, and other intelligence agencies around the world have him on their most wanted lists. Dr. Swanson entertained this as a compliment to his work and rarely exposed himself to the public in general. He enjoyed the popularity and sent emails from different locations around the world to the intelligence community, advising them of his plans.

Dr. Swanson stared at a limp body in the wooden chair contemplating the best possible combination of treatments to produce the greatest results in the shortest amount of time.

He needed to convince this kid to kill himself at the US border in two days. He also took great pride in his work and knew that with his clientele, the faster he could complete the programming, the more money he made.

He liked to set and break his own time records for this process. Dr. Swanson was vain and protective of his work, refusing to train anyone else on his methods. It kept him alive and protected by his clients, which consisted of the most ruthless terrorist groups and drug cartels in the world.

Dr. Swanson pulls back the curtain to reveal all his instruments, spread out on a sterile table. He checks the straps on the chair to make sure his patient was firmly secured but comfortable. Then he reached into his black doctor's bag and pulled out a cassette tape and put it into the player on the table, setting the volume high enough for him to enjoy Beethoven's Fifth Symphony in C Minor.

Then he went to work in precise methodical fashion, first placing duct tape over his victims' mouth, injecting him with the serum to bring him around. A few minutes passed, and the doctor noticed Spinner's eyes opening and his body twitching.

Politely, he explained to the panicking victim; "Please, Mr. Smith, don't waste your energy trying to get out of the chair. I assure you escape is impossible. You are here to perform a very special and important mission for both my client and me."

"By the way, my name is Dr. Archibald Swanson, and I must say it is a pleasure to meet you. I'm very excited about my work, and I hope you will get a bang out of it as well."

The doctor walks over to Spinner, who was more awake now and trying desperately to squirm out of the chair. He tightened the head restraint so the victim couldn't move his head from side to side. Attached to the head restraint were two little arms that the doctor pulled down over Spinner's eyes. He pulled them down and positioned them right in front of the victim's eyes. Then he opened the small, vicelike brackets at the end of each of the arms and clamped them down on the young man's eyelids.

"There you are. That wasn't so bad, was it?" Dr. Swanson asked as he starts to turn the hand screws on each of the brackets. Spinner's eyelids were spread apart farther and farther until his eyes started to tear up, and he moaned in pain through the duct tape on his mouth.

"Oh, stop it! Mr. Smith, I thought you would have more tolerance than that. You disappoint me. You're going to make this too easy."

The doctor continued until there was just a slight trickle of blood dripping from each eye. Then he nodded. "There! That's perfect! Can you see okay, Spinner?" Looking into his eyes, he whispered, "That damn duct tape makes it hard for you to speak, doesn't it? I'll tell you what, just blink your eyes once for yes and twice for no." He broke out into a sick laugh. Then he turned on the projector that flashed a picture onto the curtain behind him.

"I'm just kidding, son. You don't have to blink if you don't want to. I'm sure it's a little uncomfortable for you. Now I want you to look…very carefully…at the woman's hand on the screen behind me. Are you focusing on the hand, Mr. Smith? Good! See how her hand has a U-shaped diamond ring on the index finger. Squeeze my hand once for yes if you see it." Dr. Swanson held his left hand, and he squeezed it once. "Good! Now look very carefully at the way the hand is positioned. Notice how the middle finger is pressed into the center of her palm?" Again, Spinner squeezed his hand once. "Excellent!"

"Now, Spinner, when you get to the US border in your truck, you will find a shotgun under your seat. You will act very suspiciously to draw attention to yourself. You will basically freak out. Then when you see this hand just like this, you'll reach under your seat, take out the shotgun, and blow your head off. Do you understand… Mr. Smith?" This time, he felt two squeezes on his hand.

"Uh-oh! What we have here is a failure to communicate. Let me help you so you understand that refusing to comply is not an option for you."

Dr. Swanson got up and walked over to the big, sterile table, grabbed a drill, and a small electronic chip. He returned and showed the chip to his victim. "Spinner, this is my yes box. Once it is inside your head, you will beg me to take it out. But I won't. The only way to remove it is for you to open your own skull and pull it out yourself. No one else can do it for you. The shotgun is your tool to extract the yes box."

Then he started drilling a hole into the top of Spinner's head. Tears flowed like water through a faucet from his open eyes, and his screams were muffled by the duct tape covering his mouth. Phlegm was dripping from his nostrils and made his breathing fast and labored. His whole body quivered as he heard and felt the drill grinding through his skull. The pain was excruciating, and it seemed to last forever.

Dr. Swanson stopped drilling after a minute and looked at his drill bit. "Either this drill bit is dull or you have one of the thickest craniums I've ever seen." Then he went back to drilling. Another minute passed, and you can smell the burning bone and flesh. Finally, he breaks through the skull and ran the bit back and forth to smooth off the bone spurs. Blood was coursing down the side of Spinner's head and neck onto his shirt.

"All done…" he announced as he held the drill in front of his face and looked intently at the dripping mixture of bone, skin, and blood. "You know, Mr. Smith, I own you an apology. This drill was on low speed. It wasn't the drill bit at all. Oh well, we're through now."

The doctor walked over to the table again and dropped the bloody drill into a tray, rolled out a monitor with a long, narrow fiber optic tube attached to it, placed it next to the chair so his patient could see the monitor, inserted the tube into the hole on top of his patient's head, and a picture of Spinner's brain showed on the display.

"You see here, Mr. Smith, this little section of the brain works your saliva glands. If I push here…" He did, and at the same time, he quickly reached over and loosened the tape over his mouth. Spinner's mouth dropped open, his tongue rolled to one side, and saliva started pouring from his mouth. "Oops! See, it makes you salivate, and this section right here is your urinary or bladder control."

He pushed against it, and Spinner quivered in the chair and wet himself. "My bad, sorry… I don't know why I did that. You've been an excellent patient."

Spinner yelled out, "You fucking asshole, I'm going to kill you. Help…help!"

The doctor put a fresh piece of tape over his mouth and slapped him as hard as he could. "Ouch! Look what you made me do! You idiot, I hurt my hand. Don't you fucking realize I'm in your head? Do you have any idea what I can do to you in here?" He rubbed his hand a moment, looking at his victim, and continued.

"Now look at the screen—this is the section I want." He paused and lined up the little black chip, which had four barbs protruding from it. "It's your pain center, and by placing this little electronic trigger box right…here…" He shoved the box into place, and Spinner cringed with pain, every muscle started to constrict, and he went into convulsions. His eyes rolled back into his head.

Dr. Swanson walked calmly over to the table and grabbed a small transmitter box. He paused for a moment, admiring the distorted body of his patient, then pushed a button. "Sorry…my fault… I didn't know it was on. At least you know now why it's so important for you to get that box out of your head, right?"

The doctor walked over to him and grabbed his hand. "You do want to get that box out of your head, don't you?" Spinner squeezed the doctor's hand once with all his strength. "Good! You see, now we're starting to communicate." He removed the tube from his head and rolled the unit back behind the curtain. Then he picked up a needle and thread and walked back to suture the hole in his scalp.

Looking curiously at the top of Spinner's head, the doctor said, "Oh, damn, you know what I forgot in all my haste? I forgot to make a plug for the hole in your head. I'm sorry, Mr. Smith. Hold still for a second." Then the doctor looked at Spinner strapped to the chair and chuckled again. "My bad! What was I thinking? Anyway, I do have an idea that may work to that hole in your head."

Standing in front of Spinner, he reached into his pocket, pulled out a package of chewing gum, unwrapped a piece, put it in his mouth, and started to chew. As he put the pack of gum back into his pocket, he looked at Spinner, pulled it out again, and offered him a piece. "Where are my manners? Would you like a piece of gum?" Spinner looked away. "Okay, maybe later then." And he put the pack of gum into his pocket.

The doctor chewed quickly, looking up at the ceiling as if trying to sense the right texture. He pulled the gum out, looked at it, chewed it a little more, then pulled it out again. "Alrighty, this looks perfect." He leaned over to the top of Spinner's head and mashed the gum into the hole. He pulled the needle and thread out and sutured the scalp. He looked around for the scissors and realized he left them on the table. He shrugged his shoulders and bite off the extra thread with his teeth and spits it on the ground. He cleaned up the blood off his patient's face and neck.

"Well, Mr. Smith, that was the hard part. The rest is easy. Please stay focused on the woman's hand displayed on the curtain behind me." The doctor turned the music up a little more. Then he pulled his chair over next to Spinner and whispered into his ear, "You are getting sleepy, Mr. Smith, with each flash of the strobe light…" He looked over and noticed the strobe light wasn't on. He reached over and flipped the switch. "With each flash of light, you are feeling more and more relaxed. You want to rest, you need to rest, your body aches to rest. I will help you relax." Dr. Swanson slowly loosened the eyelid holders and pushed them back up over Spinner's head. Then he pushed a needle into his arm, and instantly, he started relaxing.

He unfastened the leg restraints and removed Spinner's soaked boxer shorts and jeans. He sprayed down the garments with an odor neutralizer and placed them in a dryer next to the wall. Then he cleaned the urine in the chair with white cotton towels and tossed them into a bio-waste container. He sprayed his genitals and legs with the same spray bottle. Then he used a handheld blow dryer to dry him off. He grabbed a bottle of powder and sprinkled it around.

After a few minutes, he removed the duct tape from his mouth. "You are relaxed, Mr. Smith. As the light flickers faster and faster, you fall deeper and deeper into a relaxing place. Think of a quiet brook in the forest, birds singing, and white, billowing clouds dancing across a blue sky. Are you there yet?" Spinner responded by squeezing the doctor's hand once. "Good, do you remember the box?" He squeezed again. "Do you want to remove the box?" Again, he squeezed the doctor's hand once. "Then you know what to do, don't you?" Another affirmative squeeze. "When I count to three, you will remember nothing until you see the lady's hand. Then you will do exactly what you need to do to get rid of that awful box. Now I want you to simply relax and rest for a while."

Dr. Swanson undid the headgear and finished cleaning his patient up. Then he released the rest of the straps and turned off the projector and strobe. He moved all the instruments behind the black curtain and called for Eric and Butch on the intercom to come down and get the package for Mr. Gilmore.

While waiting, he took off his lab coat and took the cassette tape out of the player. He sat down next to Spinner and told him, "I want you to count backward from 100 slowly. When you get to 10, your eyes will open, and when you get to 0, you will be wide awake, feeling refreshed. You will be in a hurry to get back to the Civic Center for your bus trip to Mexico. You'll remember nothing that happened to you here today. When you notice the stitches in your head, you'll remember bumping your head on a countertop when you tripped."

Eric and Butch entered the room and walked over to the chair and grabbed Spinner. "You sick fuck!" Eric yelled when he noticed that Spinner has no underwear on.

"He wet himself, you stupid moron," Dr. Swanson replied and tossed his boxers and jeans to them. "Now get him dressed quickly—he's counting." Butch held him up while Eric dressed. Then they took him up the stairs and into the office and sat him down on a soft, smooth leather sofa.

Eric sat down in a black leather chair behind a huge, dark-stained oak desk, and Butch stood next to him with his back to the bookcase. They waited a couple of minutes until Spinner's eyes opened, and then they started discussing their plans to expand the club as though they had been discussing it with Spinner all the while.

Spinner came around and looked at his watch. "Hey, guys, thanks for the tour, but I got to get going. I need to be at the Phoenix Civic Center in a half an hour."

"That's cool, Spinner. It was nice having an opportunity to share our plans for the club with a friend of Palmer's," Eric said as he got up and walked to the door. "If you get a chance to come by later, we'll show you the blueprints of the new club." He opened the door to the main entrance of the club.

Spinner responded, "That would be cool! I'd like that a lot—later!" He continued outside, where Mr. Gilmore's car was waiting with the back door open.

Palmer was already in the car, waving him to hurry up and look-ing at his watch. He got in and closed the door behind him. "Thanks, Mr. Gilmore. That was cool."

"You're very welcome. It was my pleasure. Had no idea you were going to spend so much time talking shop with Eric and Butch, though." He motioned to the driver to take off, and they headed back to the Civic Center.

CHAPTER 14

Porte Sarah

As Palmer and Spinner entered the north side parking lot, every-one was on the bus except James and Sheila, who were talking at the bottom of the steps. The bus was a nice two-tone charter diesel with finely decorated seats and flushing toilets in the back. Up front behind the driver, where most Charter buses had the first row of seats, was a built-in cooler and dry storage unit filled with snacks, soda, and water. Spinner grabbed his backpack and skateboard and ran to the bus.

The bus driver, Clinton Waiver, was dressed in a Charter uniform and checked off the last three names on the list. He took Spinner's backpack and skateboard and put them in the under-coach storage compartment and locked it down then waved his clipboard at the loiterers, James, and Sheila, as they finally boarded the bus.

Clinton waved to Mr. Gilmore, who was standing next to his limousine as he boarded the bus and closed the door. The airbrakes hissed, and the bus headed out of the parking lot, right on time.

The bus headed toward the Interstate 10 freeway. Clinton wel-comed everyone onboard and briefed everyone on the time of the trip. "Welcome to Cheyenne Charter Service, folks. My name is Clinton Waiver. I'm your driver for this trip. We will arrive at our destination of Porte Sarah in Southern Mexico in about twenty-seven hours or somewhere around 5:00 p.m. Friday. We will make a couple of comfort stops along the way for you to freshen up. Meanwhile, there are snacks, sandwiches, sodas, and water up here behind me in the coolers. We have two toilets in the back of the bus."

Clinton stopped talking for a moment while he got onto the freeway heading south toward Tucson. Then he got back on the speaker and advised, "This bus is state of the art, designed and built by Northbrook Carriage Company in Portsmouth, New Hampshire. If there is an emergency back there, all you must do is pull on the yellow cord that runs the full length of the bus, located above your window.

"I will receive a light and bell up here and will pull over when it is safe to do so. If an emergency exists while in the toilet, there is a clearly marked yellow button to push. Thank you for using Cheyenne Charter, and let me know if there's anything I can do to make your trip more comfortable."

At 5:00 p.m. on Friday afternoon, Clinton pulled the bus into the small, one-street town of Porte Sarah, Mexico, delicately maneuvering the bus through the narrow main street. The town was built on a small plateau that broke up the steep, mountainous terrain that surrounded it.

The forest was very dense with several shades of green blending together and just a splash of reds, yellows, and blues from blossoming plant life.

The town itself showed little signs of habitation, and the buildings were made of old, weathered wood covered with several coats of peeling paint. The dirt roads and paths between the buildings were made up of reddish clay. The one wooden street sign displayed the Spanish word for "Main Street."

Although it was only 5:00 p.m., the sun was behind the mountains, making it appear as though there were only a few minutes of daylight left. A short distance west of the town was a large, reddish mound of clay supporting a beautiful large, pristine white statue of Our Lady of Guadalupe. She faced east overlooking the town as though she were ever watchful for danger. The sun setting behind her cast a large shadow over the town center.

When glancing at the statue from her shadow, you saw a kaleidoscope of pale yellow, orange, red, blue, and white beams shooting onto the distant horizon and into the cobalt blue sky. Everyone on the bus sighed at the image. Clinton picked up the microphone and announced, "Absolutely breathtaking, isn't it? That is one of the most beautiful images in all of Mexico. She stands there as a constant reminder of her divine message of mercy.

"You see, this small town of Porte Sarah was named after Sarah Marie Diego, the daughter of Juan Diego, who was a simple peasant, so obedient to God that he was selected to present a miracle to the world." Clinton continued to maneuver the bus and set the microphone down in its slot.

"Well? What happened?" yelled a young twenty-one-year-old dancer from South Phoenix named Terri Thornton.

Everyone else on the bus echoed her interest too, so Clinton got back on the speaker and continued, "Okay, you asked for it! Give me a minute to put it together, and I will tell the story as best I can." After a moment of quiet reflection, he continued the story.

"On December 9, 1831, in a field on a barren hill called Tepeyac near present-day Mexico City, a beautiful lady appeared to a simple peasant, Juan Diego. The impact of their humble exchange would change the history of Mexico—its reverberations were felt throughout the world. The beautiful lady was the Holy Mother of God, the Virgin of Guadalupe who asked that a shrine be erected on the remote hill of Tepeyac, where she could love, help, and protect her people. She told Juan Diego and through him all people, 'I am your merciful mother, the merciful mother of all of you who live united in this land, of all mankind, of all who love me, of all who cry out for me, of all who seek me, and of those who have confidence in me.'

"Reluctant but obedient, Juan Diego took the lady's message to Mexico City to the Bishop Zumárraga, who asked for a sign to verify Mary's request. When Juan Diego returned to the Blessed Mother with the Bishop's request for a sign, she told him to go to the top of the hill and gather flowers and wrap them in his cloak (tilma) and present them to the bishop.

"It was on December 12, 1831, that the famous event occurred. Juan Diego took the flowers, still wrapped in his peasant's cloak, to present to the bishop. When Juan Diego opened his cloak, hundreds of beautiful Castilian roses tumbled out onto the floor, revealing a large portrait of the Lady of Tepeyac Hill. With the bishop's help, the astonished Juan Diego opened his tilma and held it up so he could see it. The bishop knelt down and gazed with wonder into the Blessed Mother's eyes as she looked down at him with a tender expression of a merciful mother's love."

Everyone on the bus was speechless as Clinton finished telling the story. The bus finally reached its destination at the end of town. The quaint little inn called Sarah's Place greeted them as the bus came to stop. Clinton added as they were departing the bus, "The original portrait on the original cactus fiber tilma still exists today, miraculously surviving over 250 years. On fabric that normally disintegrates after twenty or thirty years. It can be seen today in all its original beauty in the majestic shrine built as the Blessed Virgin requested on Tepeyac Hill near Mexico City. Folks, it's worth the trip just to view this marvelous work. And since I have most of you hooked, let me share this with you quickly.

"In 1955, an amazing discovery was confirmed by an investigative committee examining the tilma. Reflected in the eyes of the Blessed Virgin are three figures. They can be seen quite clearly as in any eyes with distortion caused by the curvature of the cornea. One figure is larger and bears a distinct resemblance to the earliest portraits of Juan Diego. The other two are less clear but are presumed to be Bishop Zumárraga and the interpreter. It is like having an actual photographic image of scene as it happened."

After a quiet and reverent pause, Clinton jumped to his feet and hustled down the steps of the bus. "We are here, folks. Please don't leave any of your belongings in the bus." He opened the under-coach storage doors and started handing out the bags.

Everyone secured their bags and stood together in front of the inn. Javier Garcia hustled out to the porch and yelled, "Welcome to Sarah's Place! My name is Javier Garcia. I'm your host during your stay with us. Please come inside, and we will get you settled into your rooms. Dinner will be served in about an hour on the back patio." Then he motioned to everyone to come in.

They entered the lobby of the inn, which is decorated with colorful throw rugs and a huge candle-burning chandelier that hung from a large arched wooden beam in the center of the ceiling. There were several old wood and wicker ceiling fans hanging everywhere you looked. There was an old open wooden staircase that ascended from either side of the lobby to the rooms upstairs.

You could see straight through the lobby to the covered patio in the rear of the inn. Servants were covering the tables with white linen tablecloths. Looking further beyond the patio was a panoramic view of the jungle valley behind the inn.

"Please, everyone, sign in on our guest register as I call your name, and I will have Guillermo and Sanchez take you to your rooms." Guillermo and Sanchez were wearing brightly colored shirts, accenting their dark-tanned skin with white ankle-length pants. They stepped forward and nodded as they heard their names called out.

Javier began to call the names of the people that would be rooming together. "Hank Jones and William Front?" As the two stepped forward and signed the register, Guillermo walked forward, took the key to room 1, and led them up the stairs to their room.

"Shelly Williams and Debra Carr!" The same thing was repeated. This time, Sanchez grabbed the room key and escorted them up the stairs to their room. Guillermo returned, and the process repeated until all rooms had been assigned.

James Evans and Rod (Spinner) Smith were assigned to the same room, number 6. Sherri Cline and Terri Thornton were assigned to room number 7. James looked at Sherri as her room was assigned and smiled, wiggled his eyebrows, and turned his head to one side. Sherri returned his glance with a smile and slowly brushed her hair over her ear.

James and Spinner entered their room together and glanced around. It was very simple and sparsely furnished, only the basics—two single beds covered with a brightly colored blanket. There was one wooden chest of drawers with a single mirror sitting next to the window against the back wall. There was a small wooden nightstand next to each bed with a small lamp on them.

On the chest of drawers was a white porcelain bowl and a pitcher filled with water for freshening up. Behind the door was a round table and two chairs. There was also a clothing rack behind the table, with a few hangers on it for shirts and pants.

Spinner's jaw dropped as he looked around the small room then asked James, "Where's the TV?"

James, seeing Spinner's disappointment, said, "Probably in the TV room or lounge." He didn't have the heart to tell him there wasn't any TV here. It was better he'd find out by himself.

They agreed to leave what they had brought with them in their bags since they would be leaving in the morning. It made no sense to unpack and then pack up again. Spinner asked, "Where's the bathroom, man?"

James responded politely, "Probably by the lobby. We're out in the boonies, you know."

James patted Spinner on the shoulder, hung his head, and sighed. "It ain't Phoenix, man, but we'll get through this. Trust me!" He tried to maintain his composure and burst out laughing.

Spinner agreed and perked up a little bit. "Yeah, you're right. It's only for a few hours. I can deal with it. After all, we're on a road trip…man."

The dinner bell rang, and everyone started heading down to the patio. The patio was lit by a few strings of lights crossing the ceiling and candles at each table, surrounded by a floral arrangement. The plates and silverware were old but clean. Each table had a basket of fresh-baked bread and tortillas.

The ambience was romantic and comfortable. There were no formal seating arrangements, and everyone started to fill tables randomly. James saw Sherri with two open seats at her table and noticed her wave off a young gentleman as he apparently attempted to sit at the table.

He tugged on Spinner's shirtsleeve and said, "Let's sit over there," leading him toward the table. As he stepped up to the table, he asked if the seats were taken.

Sherri responded, "No, not yet…why? Do you two need a place to sit? I see lots of open tables over there." She pointed to some table at the end of the patio.

"Well, I see them too, but the flowers are prettier at this table." Then he sat down.

Spinner, still standing, said, "Hey! James, those tables are okay. They got the same shit on all the tables, man."

James yanked Spinner into the seat next to him. "This is fine right here, buddy!"

Ladies dressed in colorful skirts started bringing in the plates of food to each of the tables. Everyone was enjoying themselves at the tables. You could overhear conversations covering the trip, where they were from, and why they decided to make the journey. It was very comfortable.

Shortly into the dessert, a short Mexican man walked in with a taller man carrying several envelopes. The Mexican fellow said, "Hello, my name is Miguel Ochez. I have the keys to the vehicles you'll be driving. As I call out your name, raise your arm, and my assistant Bill Ortega will bring the keys to you."

After passing out all the keys to each of the drivers, Miguel informed everyone that the vehicles were parked in a field about a ¾-mile walk south of town. He explained, "We will gather together at 6:30 a.m. in front of the inn. My friend Bill Ortega and another couple, Juan and Maria Sanchez, will escort you to where your vehicles are located.

"Once everyone is ready to go, Juan and Maria will lead you to the main highway in their 2000 Hyundai wagon—it's beige in color. Two other gentlemen, Amid and Issar Asher, will follow the group in a white 2001 GMC Suburban. They are driving the chase car in case anyone has a problem along the way. If you have a problem, pull to the side of the road and raise your hood and trunk.

"Are there any questions?" he asked with no response. "Good, enjoy the rest of your evening and this great food. We will see you in the morning. Good evening!" Then they both left through the lobby entrance.

The noise level in the room got back to normal as everyone resumed their conversations and opened the envelopes to see what kind of car or truck they would be driving.

"What did you get, James?" Sherri asked with a small degree of excitement.

"I got the perfect truck, a 2001 Dodge Ram 1500 extended cab in red. It must be my lucky day." He read the description of the key tag. "What did you get, Sherri?"

"Let's see." She opened the envelope and pulled out the keys. "Perfect! I'll be cruising in style with my 2000 Honda Accord EX in pewter. Wow! That's better than I hoped for. You sure we have to give these back?" Everyone glanced over at Terri as she pulled out the keys to her vehicle.

"Oh, shit! Not another Pontiac. I hate Pontiacs—my father has a Pontiac. It's an old person's car. It's a 1999 Grand Prix Coupe in green. Green? Hell! I might as well go up to my room and dye my hair white," Terri snorts as she throws the keys on the table.

"No, don't do that, Terri. I like your hair the way it is," Spinner said, surprisingly serious as he reached into his envelope. "Cool! I got a 1989 Ford F-150 pickup in bright orange. Wow! That matches my hair. That's so cool!" There was a quiet pause. "Man, I hope it has a CD player."

"Did you mean that?" Terri asked, looking at Spinner.

"Yeah! I love these older trucks, man—there's so much you can do with them." He tried to avoid being put on the spot.

"No, not the truck, my hair. You really like my hair?" She continued seeking a response and threw her hair to one side with a smile.

"Well, yeah! I like the way it's blond on top there and then it kind of fades into a sandy color and then it—well, you know—stops at your head with that dark, smooth color thing you got there. It tells me that you don't really mind taking chances." He looked around the table, shyly seeking approval. Not getting the response he was hoping for, he continued, "Well… I mean…you know, like you're free-spirited…that's all."

Spinner looked away, hoping a bomb would go off or maybe a fire alarm or something to save him. Terri reached over and put her hand on his. "You are so right! I am a risk taker, and I am a free spirit. I don't like anyone telling me what to do. You're the first guy that's ever seen that in me. Thank you!"

Spinner turned back, smiled, and continued to hold Terri's hand. "Yeah! Well, I just like your hair, that's all." Terri leaned forward, putting her other hand on his. As she leaned forward, her black vest parted to either side of her breast. She was wearing a skintight, low-cut white tank top that revealed two large nipple rings with a small chain traveling between them.

Spinner pulled his hand away from her, grabbing the front of his shirt. "Holy shit! You are not going to believe this, Terri, but you and I are soulmates, girl." Terri started to pull away as Spinner pulled his shirt open and revealed to her his two round nipple rings with a chain across his chest. "We are tight, girl!" He struck a pose with his head turned to one side with big smile, holding his shirt wide open.

Terri looked at him then down at her open vest and saw her rings exposed through the thin material. She looked back at him, rubbing her rings. "We are tight, aren't we, boyfriend?"

From that moment on, those two weren't more than arm's length apart for the rest of the evening. James and Sherri were amused with the event and thought that maybe this evening would turn into something special. Quietly they glanced at each other through the night, exchanging small talk, winks, and smiles.

At about 8:30 p.m., James and Sherri excused themselves and walked out to the front porch of the inn. They found two wooden rocking chairs and pulled them together. Looking out from the porch, the night was pitch black. There were a few lights on in buildings through the town.

The black sky played host to thousands of stars as bright as fireflies. It was a moonless night, and the echoed from the forest were very odd and out of place. James heard a diesel generator chugging away somewhere in the dense forest. "Maybe that's their source of electricity around here." He answered Sherri's question before it was asked. He could see the concern on her face.

The other noises, like small arms fire, rifle bolts being pulled back, and men yelling at each other in an eastern tongue like Arabic or Hebrew, were more difficult to explain. James just appeared puzzled for a while; he had no logical answer for it.

Sherri asked James, "What's going on over there?" She pointed to some men running back and forth from an older model truck parked behind one of the small houses just at the end of town.

James looked over, squinting his eyes, trying to make out what the men looked like. They appeared to be relatively tall and were wearing wraps on their heads. As they turned back toward the house, with their faces in the light, he could just make out each had dark facial hair and one had a gun hanging from a strap on his shoulder.

He could make out that they were the source of the Arabic or Hebrew being spoken. James sat back in the rocker and held Sherri's hand. "I don't know, maybe a hunting trip?" he said casually. He didn't want to alarm Sherri, and he thought to himself, *It has nothing to do with us.*

The night was filled with the chirping of geckos, crickets, tree frogs, and bats feeding on the abundance of fat, juicy insects that had now turned their attention to them, no doubt, because they were fair-skinned Anglos sitting under one of the few lights in town—veritable white meat, red-blooded buggy smorgasbord.

As they jumped up to head back inside, James glanced over once more to where the men were unloading the truck. He noticed that two of the men had also noticed them and moved closer for a better look at who was checking them out. James quickly turned and escorted Sherri inside, hoping that they were too far away to get a good look.

Inside the inn, James stopped Sherri and recommended, "Let's grab a few of those beers," pointing to the cooler by the staircase in the lobby, "and head up to my room. We can kick back, relax, and get to know each other a little more. What about it?" Sherri agrees but wanted to let Terri know and headed out to the patio.

James grabbed the beers and bounded up the staircase two at a time, whistling "Yankee Doodle Dandy" all the way up.

Sherri found Terri and Spinner kissing at the table and sat down, pretending not to notice. After a minute went by, she coughed politely. Terri opened her eyes and glanced over at Sherri. "Oh, hey! What's up? Where's James?"

Sherri leaned over and whispered to Terri.

"Great! That means you and I get my room, tiger." Terri went back to kissing Spinner.

Sherri entered the room and noticed that the light in the room was very dim. "Okay, big guy! What's going on here? Just how well did you want to get to know me?"

James smiled. "Well, I was hoping that we could talk a minute first." Then he laughed and explained that the lights were dim because they were all running off one generator.

James walked over to her near the door and handed her a beer, put his arms around her, and kissed her passionately. While embracing, he put his foot on the door and shut it.

The breakfast bell started clanging at 4:45 a.m. James woke up with Sherri in his arms on the bed; both were snuggled tightly together with no clothes on. He smiled and gave her a kiss on the cheek as she woke; she stretched, wrapping both arms around his neck, and kissed him with a little more passion.

They both jumped up, slipped on their clothes, and headed out the door to the showers. Passing Sherri's room, they knocked on the door, and it opened. Both Terri and Spinner were naked and sprawled out on the bed. The chains to their nipple rings were hooked to each other.

James and Sherri couldn't hold back their laughter. Terri and Spinner sat up in bed together and looked at them dumbfounded. "What?" Spinner asked as he looked over at Terri struggling with one of the hooks.

James said, "Breakfast! We'll see you down there."

Sherri turned to leave, and James looked at Spinner and tapped his nose. "It's a good thing to don't have one in your nose."

Spinner mimicked him by tapping his nose and then shot him a big, dumb smile and fell back into bed, bringing Terri back down on top of him. "Yeah! You're right, buddy."

Everyone hit the showers at the same time. There were only three showers on each side, so it was very quick showers for everybody. As people got into the showers, they screamed. Hot water was a luxury not available at this hotel.

About a half an hour after the breakfast bell rang, most were at the tables again eating and talking about the pending trip. James, Sherri, Terri, and Spinner ate together again, explaining that they didn't know what happened; they just ended up that way. No one was complaining.

Just as they finished breakfast, a glimmer of the sun started to touch the top of the distant forest. The sky started to turn blue again, and the colors of the jungle were coming alive. It was very quiet and peaceful except for the clattering of the dishes and some table talk.

It was about 6:15 a.m. now, and Bill Ortega walked into the front lobby of the inn. He headed straight for the checkout counter and putted out a handful of money and started to count it out for the innkeeper. After completing the count, he nodded and talked to him in Spanish and then thanked him in English for taking care of everyone.

He turned and smiled at everyone, still hustling to get their belongings together. "Chop! Chop! Everyone, please hurry. We have a long drive ahead of us, and it's very important we get started soon." Then he turned to the innkeeper again and offered a second thank-you and walked out of the door. They finally assembled outside in front of the inn. Strange, though, there was no activity in the town except for them—no dogs barking, no children playing, no women doing their chores, and no shops open. It was desolate; other than the activity that they were creating, there was nothing.

CHAPTER 15

Road Trip!

Bill got everyone together in front of him and greeted them, "Good morning, everybody! I trust you all slept well. If you would please pull out your keys and look at the tabs. They are all numbered 1 through 28. I want you all to form a line in the order of your number please, starting here with the number 1."

After several minutes of scuffling forward and back, the line was finally formed. Then Bill asked them to sound off with their numbers. There was a less-than-enthusiastic banter of number calling until it reached the end of the line. Bill then raised his hand and said, "Everyone, please stay close and follow me. Stay to the center of the path, and don't wander into the jungle for any reason. There are snakes everywhere, and some of them are very poisonous."

Off they went single-file down the reddish clay road heading south out of town. The jungle landscape was absolutely breathtaking. They air was crisp and clean. The sky was blue, and the sun had finally joined them.

Half a mile into the walk a, volley of gunfire could be heard very close by. Bill shouted to everyone; "Don't worry, folks—those are the good guys keeping the riffraff away. We've had these cars and trucks out here for a week or so, and word has gotten out to some unscrupulous people. The gun fire you hear is our boys warning any would-be thieves to keep their distance."

Finally, they arrived at a flat field of reddish clay, which cut into the jungle just off the road to the right. All the vehicles were parked in a swerving line front bumper to back bumper. The field was enclosed with a wooden fence. Razor wire was on top and woven in and out of the horizontal crossbars.

At each corner of the small compound, there was a tower built high enough to overlook the compound and surrounding jungle. Each tower was host to an armed guard with a rifle. As they approached, there was another burst of gunfire into the air.

Bill stopped everyone at the compound gate. "Okay, everyone, the cars and trucks are parked in numerical order, starting here with 1 and heading back to the end of the line and number 28. Please find your vehicle and start it up. Once everyone has their vehicle running, we will move out of the compound in the order you are in.

"This is the lead car with Juan and Maria Sanchez in it. The 2000 beige Elantra wagon is parked next to the gate. And way back there in the back are Amid and Issar Asher—they will be driving the chase car. Remember, if you have any mechanical problems or have an emergency, pull to the side of the road and raise your trunk and hood. They will stop to assist you and will radio the lead car and myself to pull over and wait. If the lead car pulls off the road, then everyone pulls off the road. Understand? Okay, go find your car and start it."

The blissful quiet of the dense jungle was violently interrupted by the revving engines. Bill looked around to see if there were any problems. Convinced all was fine, he motioned to the lead car to head out. Off they roared like a motorized wagon train heading back toward the peaceful town of Porte Sarah.

James was driving his truck in position 26. He was one of the last vehicles out of the compound. He heard a loud roar coming from the trees of the jungle behind him. He adjusted his rearview mirror to see what the commotion was all about as several volleys of gunfire ripped out into the jungle.

Behind him, he saw something in the trees like a pond of water but it was sitting vertical against the forest trees behind the back fence. The noise got louder like a tornado about to touch down. A large, dark, hideous shape emerged from the clear, water-like, upright oval.

James turned to look at what was going on behind him, not believing the reflection in the mirror was real. He noticed a large figure about ten or maybe twelve feet tall step out of the watery wall. A continuous burst of gunfire raped the jungle leaves behind the figure. It walked forward, roaring at the men firing at it.

James stopped his truck. The car behind him honked his horn then pulled around him and sped up the road. Amid and Issar Asher jumped out of the white suburban and rattled off an Uzi and shotgun at the creature. They screamed out something in Arabic, and the only thing James could understand was *Allah*.

The beast was black as night. It had two horns protruding from its skull, eyes that glowed crimson red, and skeletal hands and feet. It spoke to the men as they fired at it. It spoke to them in Spanish, and James couldn't understand what was being said. But every time it spoke, the earth trembled, and the glass in his truck vibrated violently.

Some of the men closest to it fell to the ground, grasping their ears and throats, screaming and wreathing in pain. It roared again, and another round of bullets split the jungle leaves behind the monster. James couldn't comprehend what he was witnessing. He started instinctively making the sign of the cross and reciting the Lord's Prayer. This was as natural to him as breathing. James was raised Catholic though he had stopped attending mass after Vietnam.

The great beast was about to rip the men to pieces when a large crimson hand reached through the portal and yanked the monster back into it. As quickly as it started, it was over; the creature vanished back into the window so fast that he left his arm lying on the ground. Before the portal disappeared, the huge bloody hand popped out and felt around on the ground until it secured the monster's arm. It tossed it into the air like one would do with a lucky coin, and it quickly and returned to the portal. The gunfire ceased, and the men on the ground started to recover.

Still dazed, he made the sign of the cross again and thanked God for protecting them. James could hear a horn blowing frantically; he turned to see what the noise was all about. It was the men in the white Suburban waving him on. James took off down the road toward the town of Porte Sarah.

James slowed down when he reached the town, thinking there might be people on the narrow street. But the town was just as empty as when they left it this morning. As he drove through the middle of the town, he looked up to his left at the statue of Our Lady of Guadalupe.

Her face was beautiful as it stared down at him, offering a comforting smile. Just a little north of the town, he joined the rest of the convoy and quickly jockeyed back to his twenty-sixth position.

He couldn't stop thinking how beautiful Mary looked as he drove through the town. The morning sun made her look radiant. Then he realized that something of a miracle just occurred, and he didn't even realize it.

Last night, when they drove into Porte Sarah, Our Lady was looking up over the town with her arm stretched up over her head. When he dropped through the town this morning, after that incident, she was looking down at him with a smile, and her arms were down and forward.

James was so excited that he started dancing in the seat of his truck. His own miracle witnessed in the town of Porte Sarah. He no longer had any second thoughts about the trip. He started to believe that the entire reason was on this trip was to strengthen his weakening faith. And the reason he was allowed to witness a demon and a miracle in the same day was to offer a testimony of the battle between good and evil in society today.

It was 11:37 p.m. when they hit the US border checkpoint. Most of the vehicles had already made it through and were waiting half a mile up the road in another open field. As James pulled through, he happened to glance to his left and noticed the '89 Ford F-150 Spinner was driving. That was strange because he was one of the first ones to leave.

The night sky was clear except for a small sting of clouds highlighted faintly by a quarter moon. The clouds were dark orange, like the color of Spinner's hair, and shaped like three sixes side by side. *Very strange!* James thought to himself. They were a striking contrast—the clear, black sky studded with twinkling stars.

On the side of the road just past the checkpoint was the beige Elantra station wagon. Juan and Maria were leaning against their car, waving everyone on as they passed through their gates.

Officer Pringles stopped James and asked if he had any fruits or vegetables.

"No, sir!" he replied.

Then he asked, "Have you seen anyone or anything suspicious during my visit?"

He replied no again.

James thought no one would believe him if he told them about the monster, so why say anything? Officer Pringles walked around the truck and checked the tags, the tail pipe, and looked under the hood. Then he tapped the side of the truck bed and motioned him on. "Thank you, sir. Welcome back to the United States. Please drive carefully."

The distant mountains had a dark purple hue to them as the quarter moon rising cast a faint glimmer against them. The midnight shift of the border patrol was being briefed on the agenda for the evening.

Sergeant Guerrero finished briefing his night shift. "Finally, this came in earlier today from the FBI. We are on heightened alert for any suspicious activity. Bombs and heavy amounts of drugs are expected to be crossing the border through Texas and here. We need to have our guard up and question anything out of the ordinary. Now we go out there and be safe."

Officer Kathy Spray, thirteen-year veteran of the border patrol, was finishing her inspection of a 1989 Ford F150 pickup truck. Everything appeared to be normal with the truck, but the driver was acting very suspicious. He seemed very uneasy about his truck being inspected.

Unable to uncover anything during the search of the vehicle, she asked, "Mr. Smith, where are you headed?"

Spinner responded abruptly, "None of your damn business, missy."

Then she asked, "Mr. Smith, have you have anything to drink tonight?"

To which he responded, "Maybe or maybe not, why? Do you have some?"

Officer Spray motioned to her crew sergeant to come over to her gate. Then she asked Spinner to get out of the vehicle and open his door. Spinner pulled the door shut and refused to comply with her request. She pulled the door open again and repeated her command. "Mr. Smith, step out of the vehicle now." Then she turned to see where Sergeant Kramer was.

Sergeant Kramer approached the vehicle quickly and noticed the young man reaching down under his seat. He immediately put his hand on his revolver and motioned to Officer Spray to put her attention back on the man in the truck.

As she turned toward Spinner, he hit her with the butt of a sawed-off shotgun, and she fell to the ground unconscious. Sergeant Kramer drew his weapon and pointed it at Spinner as he closed the door.

Sergeant Kramer ordered the young man to take the keys out of the ignition and get out of the truck and yelled to the other agents, "He has a gun!"

Other officers responded by drawing their weapons and surrounding the old model truck. Spinner looked over at Juan and Maria Sanchez. Maria held her right hand up with the middle finger pressed into the palm and flashing a U-shaped diamond ring on her index finger. Spinner nodded.

The situation was tense for several moments as the officers scurried for safe positions around the truck, looking for the best position to take a shot if necessary and keep out of the crossfire that may erupt.

During the scuffle, they failed to notice the exchange between Maria Sanchez and Spinner. Spinner turned back, facing forward, and revved the engine on his truck several times, indicating to agents that he was about to pull away.

A hail of gunfire erupted, flattening all four tires. The truck was disabled and unable to move. This bought a few seconds for officers to figure how to get the gun.

Spinner sat calmly, as though this were part of his plan. Spinner defiantly refused to comply with officers' requests for him to get out of the truck with his hands in the air.

Without warning, Spinner put the shotgun in his mouth and pulled the trigger. A loud bang echoed in the silent air, and the windows turned red with the blood, making it impossible to see through the unbroken glass.

Sergeant Kramer and the other officers, slowly approached the truck, not knowing if another gunman was inside. Officers moved methodically and swiftly toward the truck that now sat idling with the dead gunman slumped over the wheel.

Maria and Juan Sanchez quickly got back into their station wagon. Maria reached into the glove box and pulled out a black box with a red button in the center. She held it up and pulled out the antenna then grasped the box firmly and pushed the button.

The Ford F150 ignited into a massive plume of fire and smoke. The explosion sent the border agents and debris flying. Body parts were hurled over the cars in all the lanes, and large pieces of the truck and portions of the border gate canopy sailed into the air.

Juan and Maria Sanchez waved the last of the vehicles through the empty checkpoints. James saw the fireball behind him light up the sky and felt the ground shake with the explosion.

Border agents poured out of the barracks next to the gates, quickly trying to assess the situation and stop the other vehicles from running the checkpoints. In the panic and chaos, several cars and trucks ran through the gates, not wanting to get caught in the falling debris and the fire that was starting to spread from the truck's gas tank.

Maria tossed the black box out onto the ground and nodded to Juan to take off. Juan sped away, maneuvering around the falling debris and fire that was spreading. A few of the officers, knocked to the ground by the explosion, jumped to their feet and started to control the area by putting up roadblocks.

Three of the officers, fresh out of the barracks, direct cars into a vehicle roadblock one hundred feet south of the checkpoint, preventing others from forcing their way through.

Sergeant Kramer and Sergeant Guerrero, just arriving, got on their handheld radios and called for assistance and briefed headquarters of the situation. Sergeant Guerrero put the call in for Fire and Rescue units and advised the need for an air ambulance.

A little more than a mile on Interstate 19 north of the border in the town of Nogales, everyone was waiting for the last of the caravan to show. Bill stood in the bed of his truck with a flashlight, counting heads. "Twenty-seven…twenty-eight. Okay, that's everyone. Let's get over to MO's garage and we can go home."

Only a few people knew of the border incident with Spinner. Bill wanted to keep it that way, so he rushed everyone out from the rendezvous point and got them underway.

Three miles further north on Interstate 19, they turned off the highway and headed straight to MO's garage. The lights of the compound came on as they saw the vehicles turn off the interstate and head toward the garage. They traveled about another three-fourths of a mile to the garage entrance.

Mo's garage sat in the center of a forty-acre fenced-in compound with barbed wire on top of the chain-link fence. Two bright halogen lights handed over a tin shed or guard post at the entrance of the compound.

Three men were waiting as the first of the twenty-eight vehicles pulled up. Bill pulled to one side of the shed and jumped out of his truck. He walks over to a man holding the clipboard and shook his hand and smiled. "Hey! Bobby, how are you doing? What a trip—I'm ready to crash."

"I'm doing fine, Bill. Good to see you made it here in one piece," Bobby Hutch replied as he checked out the line of cars forming outside the gate. "Manuel is inside if you want to talk to him. I'll get these vehicles taken care of." Bill patted him on the back and headed toward the garage.

Bobby bent toward the window of the first car and asked for the number on the key tag. He checked it off the sheet on his clipboard and motioned him to pull forward past the gate to the second man wearing a white shirt.

The car moved forward, while Matt Simmons lifted the drop bar for him to pass. Joe Hastings greeted the driver of the first car and had him pull out the title to the car. Then he had the driver sign the title with his signature and put it back in the glove box and pull forward into the garage.

When the first car got to the garage, a young Arab man called Charlie opened the door and rudely demanded, "Get out of the car and remove your belongings. I will take it from here." The driver complied, and Charlie pulled the car to the back of the garage and out of the double doors.

Maria Chavez invited the driver to go around to the front of the garage and wait in the lobby for the others. "There is fresh coffee and soda available while you're waiting." He thanked her and moved to the front of the garage, stretching his tired muscles.

Then it was the second car and so forth for about forty minutes until everyone had dropped off their car and assembled in the lobby of the garage.

All the cars were swiftly moved through the garage to an awaiting crew in the back. It appeared there was a full crew for each vehicle. The garage didn't look like much from the outside, but when you pulled in, wow! It was like pulling into one of the finest dealership garages anywhere. The tools that were spread around were very expensive, and these men knew how to use them. Like a pit crew changing tires in the Indianapolis 500, these young men were highly skilled and proceeded to strip the vehicles in record time, searching methodically for something.

Vehicles were being jacked up and stripped down. Doors were cut open, and packages of something were being removed. Then another crew would go to the vehicle, open the gas tank, take out several wrapped packages, and move to the next vehicle. The drivers of these vehicles had no idea that they were moving explosives into the United States.

Everyone was settled in the lobby of the garage when Bill walked in with a tall, dark-haired man. He introduced him as Manuel Hernandez, vice president in charge of property reclamation. He said that Manuel had put the whole process together for moving the vehicles from Porte Sarah to here in Nogales.

Everyone applauded. Bill said, "Manuel has to leave and get the bus ready to take us back to Phoenix, and we should be ready to go in about ten minutes." Then Manuel slipped out and headed to the side of the garage.

Terri asked Bill where Spinner was. Bill dropped his head for a moment, trying to figure out the best way to say it. "Well, you all are going to find out sooner or later, so I'm not going to beat around the bush. Mr. Smith… Bryan Smith—I mean Spinner, if that's what he's called—was caught at the checkpoint with drugs in his truck. When the officers tried to arrest him, he put up a fight, and while shooting out the tires of the truck, one of the bullets hit the gas tank, and it exploded, killing Mr. Smith and hurting a couple of the police officers."

Terri screamed, "That's a lie! He doesn't do drugs." And she collapsed on the floor, sobbing. James and Sherri ran over to comfort her. Bill put his hands on his hips and shook his head back and forth for a minute, looking at the floor.

"Listen, I promise you we will look into this incident. We have started the investigation through our company's law office already. I just got off the phone with them. The police will be investigating the incident as well. I just don't have any other information to share with you right now. I'm so very sorry."

Bill continued, "Folks, I'm just as surprised as you are. I promise that when we get back to Phoenix, we will bring you up to date with any other information that we have. For now, we need to board the bus and go home. So if you would, please follow me."

CHAPTER 16

Bus Ride to Hell!

Bill turned and walked out of the front door, and everyone follows silently in disbelief. James and Sherri picked Terri up off the floor, grabbed her things, and headed out the door to the bus.

As they turned the corner to join the others, they stopped dead in their tracks, letting go of Terri, and dropped their bags to the ground. Terri bolted into an upright position and screeched, "What the fuck is that? You have got to be kidding me. There is no way I'm getting on that piece of shit!"

Before them stood the most hideous, dilapidated, rickety old package of bolts and steel one could possibly imagine. No, it was even worse with its spray can layers of green, black, and yellow paint with the word *Charter* brushed on the side in cursive writing.

It was impossible to tell which hit the ground first, their bags or chins. Everyone stood horrified, waiting for Bill to tell them he was just kidding, and even that wouldn't make it right. But he didn't; instead, Bill walked over to the bus and tapped the side. "She's a little aged but reliable as they come."

"Folks, the other Charter bus we have for you was supposed to be here four hours ago—it never showed up. We got a call about an hour ago from Cheyenne Charter Service. Evidently, the Charter bus you were supposed to go home in threw a rod coming back. It's still somewhere in Mexico."

Everyone, included Terri, was too hurt, tired, and worn-out to put up and kind of protest. So they boarded the 1962 International chariot from hell, moaning, groaning, and some crying on the way.

James, Sherri, and Terri went straight to the back of the bus and filled up the rear seat. James figured if the bus broke in two, they would have better chance in the back near the emergency exit.

Once all were on board, there was a delay waiting for the bus driver. Bill stepped on the bus and announced, "The bus driver's name is Pops Garcia—he's been driving this bus for twenty years. He likes to be called Pops. Everyone in the company knows him."

He started to step off the bus then hesitated. Then he said, "Oh, we are also sending two of our security guards with you, Akmal and Muhammad, for the rest of the trip. You know, just in case." Then he stepped off quickly, pretending he didn't hear the desperate cries for help.

Manuel was at the back of the bus, talking to both men that would be riding with them in Arabic. James admittedly knew nothing of the Arabic language and not much more about the eastern culture. But he did know body language, and when Manuel barked at the two men and ran his finger across his throat, it wasn't to wish them a happy birthday.

The two men, Akmal and Muhammad, hurried to the front of the bus and boarded it, sitting together behind the driver's seat. A moment later, Pops strolled around the corner, heading for the bus. He was a sweet-faced, old Mexican man who looked as though he had just come from the fields.

He was a short-framed little man sporting a big sombrero with a white scarf hanging down the back of his neck. Manuel grabbed him by the arm as he passed and whispered something in Spanish. Pops nodded affirmatively, pulled his arm away, and hustled to the bus.

Pops jumped into the driver's seat and said, "Hello, I'm Pops!"

They acknowledged him as best they could by responding "Hello!"

Then Tom Porter yelled out from the middle seat. "Pappy, for every hour you shave off the trip back to Phoenix, I'll give you $100."

Pops smiled politely. "Sir, I'll get you there safe—that's all I can promise. Now for those of you who haven't ridden in a '62 classic like this one, you find seatbelts under the spare tire in the last row. They look like ropes, and you just tie it to you and something that doesn't move…okay? Okay! Here we go!"

The bus revved up and headed out the front gate down the short, bumpy road to the Interstate. The fumes from the diesel engine were heavy inside the bus. "Turn the air conditioner on, Pops!" someone yelled out.

"Sorry! No air conditioner on this bus, but it's a very nice night tonight and will cool off quickly if you let your window down." Pops responded just loud enough to be heard over the grinding noise of the antique diesel engine.

The bus slowly reached its maximum cruising speed of about sixty miles an hour. It only took about a minute to reach it on Interstate Route 10 heading north toward Tucson. The speed limit was 65. The old diesel engine vibrated the entire cab and provided a soothing hum that mixed nicely with sound of the engine. It made it very difficult to understand anything being said on the bus even from the person sitting next to you.

Most of the people on the bus struggled to make themselves comfortable for the five-hour trip to Phoenix. James tried to comfort Terri, still sobbing over the loss of Spinner, by putting her head on his shoulder and explaining loudly, "There is no way you could have known what Spinner was going through. Were all shocked about the whole thing. Just relax and rest here for a while we'll be in Phoenix soon."

Sherri smiled as she witnessed how compassionate her new beau was with others in a time of crisis. Then she pulled his free arm around her and rested her head on the other shoulder.

James looked forward at Akmal and Muhammad. They appear to be fidgeting with something in their seats or in their laps; it was hard to tell because the back of the seat was concealing their activity. He was suspicious of the two men being presented earlier as security but not wearing any sort of uniform that might identify them as such.

Roughly about fifty-five minutes into the journey home, the bus slowed and pulled off the main road and started traveling down a dirt road, leaving a wake of thick dust behind it. James yelled up to the driver, "Hey! Where the hell are you going? This isn't the way back to Phoenix!"

The two men of eastern descent shouted back, "Relax, we need to make a stop at the bus driver's sister's house—it's his niece's birthday today. It will only take a little while, and we will be back on the road in no time."

The road started to get more uncomfortable with the bumps and washboards jostling everyone around. It was a very dark night, and nothing could be seen from the bus in any direction. The other occupants of the bus voiced more objections. But the driver stayed his course with no comment, staring straight ahead. The only signs of light in the darkness were the two columns provided by the bus's headlamps that cut sharply into the pitch-black night and ended about a hundred feet in front of the bus.

James was about to voice another objection when he noticed the silhouette of a large utility vehicle traveling behind the bus with only its running lights on. He could barely make it out through the thick cloud of dust and the pitch-black night. It looked like the Suburban Amid was driving earlier as the chase car. He shook Terri and Sherri off him and whispered to them both that something wasn't right, pointing out the vehicle behind them.

James was about to walk to the front and confront the driver and security personnel when Akmal and Muhammad stood up and yelled to the people in the bus to remain calm; the accelerator on the bus was stuck and the bus driver was going to try to loosen it in a moment. Pops leaned forward for a second and then leaned back in his seat. The diesel engine started to whine with power.

They told everyone to move to the center of the bus for their own protection and that they and Pops were going to try to stop the bus with the emergency foot brake. People in the bus started to panic and gradually huddled to the center of the bus. James shifted positions with Sherri so he could be on the aisle edge of the seat in case he needed to act quickly. He noticed Pops violently shoving and pulling the automatic transmission lever forward and back until it broke off in his hand.

James yelled out, "What the fuck are you doing to the column, asshole!"

Sherri grabbed him as he started to stand up and run forward, pulling him back into the seat. "James! Don't do anything stupid—those men could hurt you."

The bus was traveling about twenty-five to thirty miles an hour, probably the best speed it could reach on this bumpy road. The two eastern fellows moved toward the front door of the bus and yelled at the bus driver, "Now!" The bus driver slammed his foot on the brakes, sending everyone in the center of the bus screaming to the floor. James and the girls slid forward into the back of the seat in front of them.

James hit his head on the corner bar, causing blood to flow from a cut over his right eye, which caused him to become disoriented for a moment. The front door flew open, and Pops jumped out of the bus along with two eastern fellows. He left the bus in gear when they jumped, and the bus began to chug along, gaining speed quickly.

He looked out the back window of the bus in time to see Pops, Akmal, and Muhammad brushing the dust off their clothes. He noticed the lights of the suburban illuminating the scene beside the bus as it continued past them. The two security guards loosened their garments, revealing their earlier discomfort; they were carrying Uzis under their garments.

Pops looked surprised at the two men as they pulled their weapons out and shot him. He fell like a sack of potatoes into the dust along the side of the road. Akmal and Muhammad jumped into the suburban as it continued to follow the bus.

Two men and a woman managed to break free from the huddled mass of bodies in the aisle and headed to the front of the bus. One of the men grabbed the wheel and tried to turn the bus off the road to slow it down. The steering wheel would not budge; it was locked into position with a lock bar. The other man tried to pull the gas pedal up away from the floor, but it too was jammed into place with another lock bar.

The woman, seeing the hopelessness of the situation, jumped out the open front door of the bus and tumbled to a stop alongside the dirt road. Then the two men who had tried to fix the problem jumped with similar results. James saw the suburban slow down and heard a burst of gunfire from the Uzis. Three bursts rang out into the otherwise silent night air. James could see the flash from the barrel of the guns as they shot each of the people that had jumped from the bus.

As the bus continued to pick up speed, it started to bounce wildly, sending people and baggage high in the air and down on top of the seat bars and to the floor again. You could hear the snapping of bones and screams of pain and horror muffle the whining engine and creaking of metal as the bus continued to romp helplessly into the desert.

James tried desperately to hold on Terri and Sherri as they escaped his grasp and were thrown into the mass of bodies in the center of the bus. He was secured, somehow, against the back of the bus by his belt; it was caught on something solid. As the bus bopped up and down, he felt a sharp jabbing in his back but remained anchored to the object.

James turned his head back to see what was happening. Behind them, the suburban was stopped with its headlights illuminating the edge of the road. The bus was heading off the road into the desert.

Suddenly, the bus was sailing into the night air. The nose of the bus dropped forward and dove into a large ravine. A lady screamed out, "Oh my god! Someone help us please." Those that were still conscious tried to grab on to solid pieces of the bus. James could only watch from his hanging vantage point impaled on the emergency exit door handle.

As the bus dropped into the ravine silently, its headlamps lit the bottom of the gorge, showing their impending doom for a moment before they hit. Then there was the horrific sound of crushing metal and broken glass. Bodies slid to the front of the bus like bags of flour stopping suddenly with a loud thud and grunt. Most of the people in the bus went silent on impact; a few were still moving and groaning.

James could feel himself lifted out of the bus as though he was being catapulted away from the vehicle. His belt was still wrapped around the emergency exit door handle. During the impact, the emergency door flew open, ripping him out of the bus. As the door met with the back of the bus James's belt gave way, and he continued his flight into the soft dirt edge of the cliff they had just traveled over. He landed in a sitting position with his back against the wall.

He could hear a few screams for help but was unable to respond. He couldn't move a muscle, the wind completely knocked out of him. He sat wheezing for oxygen. He could smell the thick stench of diesel fuel leaking from the tank. He mumbled to those that were still in the bus to get out of the bus, thinking it was going to blow up.

He glanced down to see what had cushioned the fall of the bus and noticed that this wasn't the first bus to make the bottom of this one-hundred-foot-or-so drop. Under the bus were other buses, more than he could count. James thought to himself, *How could this possibly happen to so many people and no one be aware of it?*

James heard scuffling above him; it sounded as though there were rocks falling from the top of the edge of the cliff. He could faintly make out the voices with a heavy eastern accent. He heard them counting down—one, two, and three—and another body traveled off the cliff and sailed to where the bus had landed about fifty feet away from him. Then there was another and another and still another body being tossed over the edge of the cliff.

The last body thrown over the cliff was that of the bus driver. He knew that because his sombrero sailed down above him slowly. There was silver tape on him and something black. As the body hit the back end of the bus, part of the body made it into the emergency door, and the legs dangled outside on the bumper.

Two—maybe three—seconds went by then a huge explosion; the bus blew into pieces and bodies flew everywhere. The diesel fuel erupted into a huge ball of flame and black smoke and almost reached James. The heat was so intense from the blast it singed the eyebrows on James's bleeding forehead. There was laughter from above and then the scurry of feet running away, the closing of the car doors, and he could hear the suburban speeding off.

CHAPTER 17

The Old Man

James sat there confused and stunned; he was still trying to imagine what just happened. He stared into the orange and yellow flames mixed with thick, black choking smoke that spiraled and mushroomed into the pitch-black sky. He could smell the fire as it engulfed the bus and its unfortunate remaining occupants. The burning diesel fuel, flesh, rubber, plastic, and paint presented an aroma of death that James had been exposed to before when he was in Vietnam.

The horrifying image rekindled thoughts he had buried many years ago. James began to weep uncontrollably, sensing the loss of his new soulmate, Sherri. "Why…why…why?" he screamed, looking toward the heavens for an answer. "These are good people. They didn't hurt anyone. Why would you take them like this?" He sobbed louder; his body was racked with pain. His heart was ripped in two with grief for those lost.

"Why didn't you take me? I'm the sinner here, not them. I'm the one who cursed you in the field of battle. I'm the one you really want. Why them?"

"You're asking the wrong question, young man," a voice echoed from the blackness of the night.

James, startled, looked around to see where or who the voice was coming from. "Who's there?" he said as he looked around trying to find a stick or a rock he could use to defend himself with. Seeing no one and no reply, he shouted again, "What do you want?" His voice was tensing up.

"Don't get your underwear in a bunch, young fellow. I'm not here to hurt you." Slowly a figure emerged from the blackness of the ravine floor, moving in his direction. It was an older man with a rod and a staff. He walked slowly toward James, glancing at him occasionally then dropping his head to watch where he is stepping.

The old man stood next to a body thrown from the bus during the explosion. He crouched down next to the person and blessed him, making the sign of the cross "In the name of the Father, the Son, and the Holy Spirit, I commend your soul to God Almighty. May you rest in peace."

"Who are you? A priest?" James yelled out as he witnesses him blessing the man. The old man looked up at James and put his finger to his lips, insisting on his silence. Then he stood with an image of the man that was lying dead on the ground standing before him. A transparent image in full color smiled at the old man and bowed his head. Then the transparent image ascended.

The old man walked over to within several feet of where James was sitting. "My son, I'll be with you shortly. But for now, please sit there silently and don't move." Then he turned and walked toward the bus, stopping at the bodies that were around it and duplicating his actions with each one.

The old man missed one of the bodies on the ground as he walked toward the burning bus. James yelled to the old man, thinking he had overlooked him. "You miss—" He was silenced by the old man as he turned and waved his staff.

The ground shook and cracked open near the body of the man that he had passed over. Standing over that body was the transparent image of Pops, the bus driver. He was yelling at the old man to have mercy on him. The old man ignored his plea.

Dark, hideous creatures, black as tar, climbed up from the orange-red glowing cracks. They had heads like jackals, snarling and growling as they reached out toward him. Their eyes were crimson fire, and their bodies were like black skeletons. They slithered out of the hellish cracks in the earth and surrounded Pops. He tried to fight them off and cursed at them. As they grabbed him, his colorful translucent image lost its luster of color and turned gray and then black like them. They swiftly dragged him down into the crevasse as it rumbled and closed on top of them.

After the old man had blessed the bodies around the bus, he entered the fiery blaze and was out of sight for several minutes. James thought the old man was in trouble and tried to move toward the bus to help him. He couldn't budge.

Then several transparent images of those on the bus when it caught fire began to ascend out of the flames. He saw Terri and Sherri holding hands and smiling as they ascended. They looked over at James sitting there helplessly and waved at him.

Tears flowed down James's cheek as he watched them ascend to a brilliant light in the sky. He couldn't help feeling sorry for himself, having lost the most wonderful thing that had happened to him since the war.

The old man emerged from the flames and headed toward James. He didn't have a mark on his body nor were any of his hairs or clothing singed. James was awestruck as the man approached him. The old man waved his staff again, and James was able to move and speak again.

James looks at the old man with curiosity and fear. The old man came closer until only a few feet away. "So! What is it?" he asked the old man.

"What is what?" he responded, reaching down and picking up a handful of dirt.

"What is the question I should be asking? You said earlier that I was asking the wrong question. So what is the question I should be asking?"

"Oh! That! Well, let's see if I can explain it to you simply." He spit into the dirt in his hand and rubbed his hands together, making a small pile of mud. "Open your shirt and raise your arm, James," he said as he walked closer.

James complied reluctantly at first, opening his shirt then raising his arms slowly, watching the old man come closer and rub the mud on the right of his chest and then on his forehead. James felt the pain subside and a tingling over his body where his wounds were.

"You had three broken ribs, a fractured tibia, and a cut over your eye."

The old man stood back and watched James as he began to realize that he had just healed his wounds. "Wow! How did you do that? I feel great! I can breathe again." He took an exaggerated breath and let it out.

"Again, you ask the wrong question. Instead of asking so many questions that you already have the answer to, you should be thanking him who sent me to you."

James started without thinking to ask another question. He stopped himself when he saw the old man shake his finger at him scornfully. Then as though he knew all along what he had to do, James dropped to his knees, crossing himself. "In the name of the Father, the Son, and Holy Spirit, I offer my thanks to you, O Lord, for sparing me from certain peril. Amen!"

The old man put his hand on James's shoulder and knelt beside him. He crossed himself and prayed with him. "Oh, Lord God Almighty, we offer our thanks to you as we kneel here before you as humble servants. Show to us your will this night that we may better serve thee. Strengthen us in our understanding of your love. Enlighten us to the task at hand. For all power and glory is yours forever and ever. Amen!" They make the sign of the cross and stand up together.

As they stood, a choir of angels sang, "Amen!" James looked up into the black night sky and caught a glimpse of the angels singing as they faded into the night sky.

The smoke from the burning bus tires and diesel fuel overcame James, and he fell to his knees, coughing and holding his throat. The old man waved his staff in the air, and a pillar of light surrounded them. James starts to breathe again without coughing and stood up next to the old man as he waved his staff again.

A perfect hole formed in the ground next to the bus, and the flames and smoke were sucked into the hole as though it were a huge vacuum. A moment later, the bus was completely extinguished of all flames and smoke. It toppled on its side, covering the hole.

The pillar of light vanished, and both were standing there looking at each other. James noticed that both their bodies were illuminated with a bluish glow enough to light five or six feet. It seemed to have no origin but allowed them to see each other clearly.

James asked the old man again, "If I'm not supposed to ask questions, then how am I supposed to know what I need to do? What is this task you referred to in the prayer?"

The old man looked pitifully at him and said, "You, young man, are to help launch a modern-day miracle of God's love. This miracle will unite all the nations around the world to live in peace and harmony for some time."

"Holy shit!" James slipped out.

The old man looked at him, shaking his head back and forth. "Not likely, son. You need to ease up on the cussing, tough guy. God's patience can be tested. Remember the story of Lot's wife? All she did was look in the wrong direction and she turned into a pillar of salt. I can only imagine what he has in store for those that verbally affront him."

"You're right. I'm sorry, it won't happen again!" James said to the old man and looked up to heaven.

"He's equally provoked by promises that can't be kept."

"Yeah! You're right again. I'm sorry, Lord. I can't promise I won't cuss anymore, but I promise to be more mindful of it." James looked over at the old man to see if that was okay. The old man glanced over to him and gave him an affirmative nod. James smiled and nodded several times like a schoolboy after getting the coach's approval.

The old man and James walked several yards down the ravine away from the crash site. The old man stopped and sat down on a large rock. James sat across from him on another rock. The old man started to explain what he needed to do.

"James, God has heard the cries of his people around the world for peace and harmony among man. He has heard them from the churches and synagogues in every city, state, and country. When man is given violence, he responds with prayer. When they lose a loved one, they cry out for comfort. He can no longer withhold his love.

"There is a child in New York City that is deaf and dumb. She carries in her the gift of the Lord. He has chosen her because she is selfless when others are selfish, courageous when others are fearful, strong when others are weak. She shows love to those who hate and despise her. Above all else, she has faith in the Lord God Almighty despite all her sufferings and hardships. Her name is Sarah, and on the holiday that most of the world celebrates as Christmas Eve, she will offer that gift to the world from a great cathedral."

"So what do I do?" James asked, very puzzled.

"You must ensure that she is at the great cathedral to present the gift. No place else will do. Any place else will corrupt the gift, and the world will be tormented by Satan and his demons."

"I must warn you now, every effort will be made to stop you from reaching her. Satan is very strong right now on earth. He has servants in every arena you will pass through. He has instructed them to stop you, even kill you if necessary to keep you from your task. He knows all your thoughts and especially all your weaknesses. He will use them against you."

"But why me? I'm no saint. I killed people in Vietnam. I never went to church until a couple months ago. I've never done anything great. I can't even keep a job. Why would God want me?"

"You see what I mean. Right now, Satan has placed doubt in your mind. These are his tricks, and he will use them against you. Okay! I only have a few minutes. But let me try and explain what God sees in you. Four weeks ago, you fell to his feet at St. Anne's Catholic Church in Gilbert. You begged him to forgive you. You were rent with pain and sorrow before the feet of Father Doug. Jesus said to you, 'I absolve you of all your sins. Go forth in peace, my son.'

"Your heart was whole again. The pain was lifted, and you praised God for hearing you. Your faith has made you whole son. Your peace with God has made you strong. You have been through many trials and tests. With each you have grown closer to him and stronger in your faith. Now, you must complete your mission. Unlike everyone else whom has been chosen before you, you have been granted foreknowledge of your task. Do it well!"

There was a rumbling in the earth as the ground started to shake. "Run, you're in danger! Run to the end of the ravine as fast as you can and don't look back for any reason. There you will see a white horse grazing. Extend your hand to him, and he will carry you to safety. Go now! Quickly!"

James bolted off running as fast as his legs would carry him. Looking straight ahead, he could just make out the edge of the ravine dead ahead but still quite a distance away. He could hear thundering and lightning behind him and a loud, ripping sound. But he didn't look back.

The old man sat on the rock, slapping his staff on the ground. Each time his staff hit the ground, a loud roar of thunder echoed through the small ravine, and lightning darted from side to side. The earth was boiling in front of him as though the ground were a hot, dirty bowl of liquid. When a bubble burst, another creature would spring up and start running toward the old man. He struck his staff on the ground, and a clap of thunder would drop the creature to its knees, then the lightning would ricochet off the walls and strike the creature, sending him underground again.

Over and over the old man repeated this process, trying to keep the creatures at bay. More and more creatures would surface after each attempt. The ground was boiling violently, and the numbers of creatures were doubling. The old man kept hitting the ground with his staff faster and faster

James could hear the claps getting closer and closer together and knew the old man wouldn't be able to keep that pace up much longer. He ran faster and faster with each stride. Then he saw the end of the ravine; it was close now. The claps of thunder were almost one huge, long roll of thunder now. Then suddenly, all went quiet. James stopped at the edge of the ravine and saw the horse grazing just as the old man told him.

CHAPTER 18

The Magic Horse

James was breathing heavily, struggling for more oxygen as he exited the ravine. He thought he should go back and help the old man get away from whatever was back there. But a voice in his head told him to run. Without looking back, he ran as fast as he could until he got within ten feet of the horse. The horse got spooked and start to bolt. James stopped and stretched out his right hand. The horse stopped and turned walking cautiously toward him then picked up his pace.

James sensed danger was imminent; he could feel the evil closing in on him. He stood his ground and walked slowly toward the horse to meet him halfway with his right arm still extended as the horse got close enough for him to touch it. He looked into the horse's eyes as the horse started to pull back and saw hundreds of crimson red eyes focused on him. He leaped onto the horse's back and held on tight with all his might as the horse charged off.

James buried his head into the horse's neck and interlocked his fingers into the horse's mane. The horse flew across the dark, sandy rock floor of the desert as the demons started to lose ground. James had never ridden a horse as fast as this he was like the wind traveling east.

James raced through the southeastern portion of the Arizona desert under a dark midnight sky. His white stallion galloped like the wind across the dusty desert floor. James gave no care to where his steed was taking him, only that he had successfully escaped the grasp of the evil demons so bent on his destruction.

While he rode on the back of this tireless steed, he had a vision of a beautiful young girl hanging from a tower on top of the empire state building. She was surrounded by the same demons that chased him into the desert. They were forcing her to let go of her grip on the tower, and no one in the city seemed to care. They were too busy with their mundane daily lives.

Then James shook himself conscious and back into the nightmare, which had become his reality. He knew now that his mission is to find and save the little girl in his vision. He removed his belt and strapped the horse from side to side to increase its speed. The magnificent stallion responded favorably and increased its speed across the desert floor. A cloud of dust rose behind them as the mountains passed by like a picket fence.

"Holy shit!" James screamed as he realized that the steed was traveling more quickly than his Dodge pickup ever could. He placed his hand below the horse's neck to feel its body heat and pulse. "This is unbelievable!" The horse was cool, and its pulse was steady. "Where did you find a power bar out here in the middle of the desert?" James asked, amazed at the stamina of this beautiful creature.

Just a hint of light started to peek through the darkness over the top of some nearby mountains. James noticed that the horse was traveling so quickly that the terrain was flying by in a blur. He had no sense of the direction they were traveling, only that they were heading toward the rising sun.

James could feel the power of the horse generating in his own body now. He felt invincible on this mighty steed. He strapped the horse again to see if he could go faster, and the white stallion reared his head back and took the belt out of James's hands. Holding the belt in his teeth, the belt slapped James several times as the wind caused the belt to flap like a flag.

"Ouch! Shit! Ouch! Stop it, that hurts!" The horse finally released the belt, and it sailed into the dust cloud behind them. Then James heard several screams and thuds and slaps coming from the cloud they had created. Attentively, he scanned to either side of the wall of dust thrown up behind them. He noticed the glowing red eyes of the creatures he thought they had left miles behind them.

They hadn't disappeared at all; they were traveling as fast as the horse and gaining a little. With nothing to strap the horse with,

James yelled out, "Will you please get your bony ass in gear? Those monsters are right behind us."

James thought the horse was unaware of the danger behind them. But the horse was very aware and continued forward as fast as it could, galloping toward the narrowing canyon walls. The closer they got to the narrow entrance to the canyon, the tighter it became for the charging demons.

As they got closer to the entrance into the canyon wall, the spiny tree and cactus branches became more of an obstacle. James looked forward and noticed that the entrance was barely wide enough for the horse. He knew then that the horse was very much aware of the situation and was forcing the demons to follow into the canyon to slow them down.

He dodged branches of cactus and spiny tree limbs leaning back and forth and from side to side. Faster and faster the branches hurled toward him. Some of the branches left their mark as they grazed him on the arm and shoulder. His knee captured the entire limb of a jumping cactus. James could feel the intense pain of the needles driving deeper and deeper into his leg.

The opening was just a hundred feet ahead of them, and the canyon walls were so close together James could reach out and touch them. He tucked his head tight on the side of the horse's neck and pinned his legs and arms tight against the body of the horse. As they shot through the entrance, it opened into a large sandy floor oval.

The sheer walls of the canyon shot straight into the sky above. They were smooth as polished stone. They were completely closed in on all sides except the narrow path that led them there. James was shocked, thinking that this dumb horse had just sealed both their fates.

"What the heck! The old man said you were to take us to safety, not certain death!" James and the beautiful white stallion began to gallop in circles along the canyon wall, looking for a secret exit or a passage that they had missed. Everything seemed hopeless as the guardians of hell pushed and shoved their way through the small opening into the large circular room.

They darted for the center of the canyon as they popped through the opening. The demons of Satan stood their ground in the center of the canyon, waiting for the right numbers to gather. They grunted to each other, causing some to stand fast at the entrance and allowing the rest to form together in the center. They were waiting for the right moment to attack and put an end to the chase.

James knew they were finished; he had scanned the entire canyon for a way out, and the only out was the way they came in. "Well, my good friend, we gave them a good run for their money, but I'm afraid that's all she wrote." James determined the only way to go was with a good fight.

He stopped his trusty steed against the wall farthest from the entrance. Then he stared down the beasts that wanted him so badly. They were growing in number every second, and soon there would be no room for them to move. "Well, the way I see it, we have two options. One, we sit here and let them kick our ass or we charge them and take out as many as we can before we go. I'm all for number two! Hee yaw!"

James jabbed his heels into the horse's side. The horse reared straight up and then charged to the center of the canyon. The demons reacted with a charge of their own. Just as they reached the black island of demon stench, the beautiful white stallion offered a third option and jumped high above their heads.

Then it circled around in the air, flying several feet away from the canyon wall. With each circle, the horse rose higher and higher until they finally cleared the top of the canyon. "Well, that's always an option too," James expressed aloud as they cleared the top of the canyon and flew toward the rising sun. "Mom! Look! I'm riding a flying horse and being chased by Satan's army! And you were so sure I wouldn't amount to much! Look at me now, Mommy!"

"This is unbelievable! The guys are never going to believe me. Listen, Pegasus! You think we could swing by the club on the way and have a beer with the guys? I'm buying!" James continued to talk to his gifted stallion as he headed for a nearby mountaintop.

James and his trusty companion landed on Table Rock, a tall cliff in the middle of the desert mountains of New Mexico. He dismounted the horse and stood quietly observing the beautiful sunrise.

The horse stopped there for a reason; he just didn't know why. James was relieved those hideous creatures weren't chasing them anymore.

It was called Table Rock because it looked like a giant table made of stone, cut away by the wind and rain for thousands of years. The surface was flat and grainy like fine sandpaper. There was nothing else on the mountain when they landed on it. It stretched out about 150 feet in diameter.

A voice broke James's meditation on the sunrise. "Welcome, James. I've been expecting you." James turns to find an Indian standing behind him watching the sunrise with him.

"How long have you been standing there?"

"Not long! Are you ready for the next step on your journey?" the young Indian warrior asked.

"That all depends on what the next step is, my friend," James replied.

"I understand you had some bad spirits chasing you. You're probably a little concerned about your own safety. I want you to know that is why I'm here," the young Indian warrior offered.

"A little concerned is putting it too mildly. I'm very fucking concerned about my safety. Who are you and what role you play in all this?" James inquired.

"I'm your spirit guide, and I'm here to help you become less visible to those that would harm you. I was called upon to show you how to walk in the duality of the spirit world. This will make you harder to locate by the evil that tracks you." The young Indian warrior pointed down to the floor of the desert below them.

The floor of the canyon was slowly turning black with the army of Satan's demons. "As you can see, they aren't far behind you now. But don't worry. You're on sacred ground, and they can't hurt you here. We should get started with your training." He pulled out a long pipe from his waist belt and lit it with a burning twig.

As he puffed on the pipe and handed it to James, he explained, "My name is Running Bear. I'm Navajo warrior, and I rode with the Great Chief Geronimo. I was trained to walk in the duality of the spirit world by a Great Shaman known as White Owl. We need to start by purifying the body and freeing your spirit."

James puffed on the pipe a couple times and handed it back to Running Bear. "Remove all your clothing and kneel down by the fire."

James complied with the request reluctantly. "How is any of this going to get rid of those creatures down there?"

Running Bear kneeled next to James by the fire and started to chant and rub ashes on his chest and forehead. After chanting to the spirits and rubbing ashes on him, Running Bear explained, "First, we block the evil spirits from entering your heart or mind while you are purified. Then we enter the sweat lodge over there and stay in there until you have been trained." Running Bear pointed to the opposite side of the rock where a shack had been constructed.

"The sweat lodge allows your body to purge itself of any bad feelings, and it frees your soul so that it can be trained in the ancient arts of our ancestors." Running Bear stood and walked slowly to the shack. James followed him quietly. He figured it couldn't hurt and sure didn't want to be chased all over the place by those hideous monsters.

"Ah! What the hell!" James blurted out as they approached the sweat lodge. Inside was a huge fire pit filled with red-hot coals. Around the wall of the shack was a makeshift bench to sit on and the smoke pouring out the large hole in the roof. Once they entered the shack, the door was closed and sealed from the inside.

The heat was intense, and James fought to get a breath. "Relax, James. Don't fight it. Allow your body to slow down and release its energy. You have to release yourself before you can step over," Running Bear explained as he sat unfazed by the heat. He passed James the pipe again and instructs him, "Inhale the smoke and hold it. Then release it slowly and repeat three times."

James did as he is instructed, and after the third time, he felt quite euphoric. He was no longer chanted in the Navajo language. James began to understand every word he was chanting and smiled. Finally, James drifted off into a comatose state of mind.

It was as though he was in his own dream and Running Bear was there with him. He saw a big, clear, beautiful lake smooth as glass, surrounded with tall pine trees sitting under a blue sky with white, billowy clouds floating overhead. They both were standing in the middle of the lake on top of the water facing each other.

Running Bear looked at James and stated, "Now that we have crossed into the realm of spirits, we can begin your training." James was still taking in the cool breeze and the gorgeous surroundings. "Wow! Far out, man! Is this stuff legal?"

They turned and walked to the shoreline of the lake together. On the shore a bonfire is blazing, and several medicine men were singing and dancing around it, calling on all the spirits to join them.

CHAPTER 19

Contract in New York

The Celestine Investment Group company jet landed at Sky Harbor Airport in Phoenix. The aircraft taxied into a large private hanger at the south end of the terminals. A fleet of limousines was lined up waiting for the passengers. Bob Mackey, CEO of CIG, deplaned first, followed by his entourage: Jeff Corbel, his chief of operations and right arm; Susan Cromwell, his private secretary and friend specializing in overnight stays; and Tim Porter, his lawyer and personal legal advisor specializing in corporate law.

"Hi, Bill Britt. How the hell are you? Is everything ready? I got to be out of here by noon tomorrow," Bob shouted to a tall man at the bottom of the steps.

"I'm great, sir! It's good to see you again. Everything is ready. We'll have no problem getting you out of here on time."

"Good God! I love to hear that! That's music to my ears," Bob shared with Bill as he reached the bottom of the steps and shook hands with him.

"I know, sir! That's why you have me here!" Bill responded with a smile.

"Is the meeting set up with the Smith Brothers Construction Company tonight at seven?" Bob asked as they briskly shuffled to the limousine.

"Yes, sir! All set for the Rusty Pelican at seven sharp," Bill stated as he opens the limousine door.

"Every time you open your mouth, you remind me why I pay you so much. You're the best damn events coordinator I've ever had working for me."

"As it should be, sir, as it should be! I'll see you tomorrow at 11:00 a.m.," Bill finished as he closed the door to the limousine. The rest of the team jumped into the trailing limousines, and the caravan headed to the Ritz Carlton hotel downtown.

Bob Mackey picked up the car phone and had the driver notify the other team members that there would be a strategy meeting at 5:00 p.m. in the Blue Buckle room at the hotel. Then he took a deep breath and let it out slowly. Finally, he was going to have some of the nicest buildings in New York City constructed by an outside company. He'd been dreaming up this plan for five years.

Back at the Smith Brothers Construction Company office, Bill Smith and Frank Smith were having a cup of coffee, discussing the meeting with CIG. "This is a big contract, Frank. Are we sure we got everything covered?" Bill said nervously.

"Everything is covered, and the contract is general with several options. If we get the one done on time, we have the contract for five more buildings."

"Shit, Frank, that's a lot of work. I definitely want Mike running this one. There is no room for error," Bill shared as he finished signing the contract on his desk.

"We're fine, Bill. I'm having dinner with Mike and Mary tomorrow at the Camelback Inn. I'm going to offer him a promotion to project manager and offer to pay for all moving expenses. Mike's a team player. He'll be more than happy to run this one for us," Frank responded as he collected the signed copies of the contract.

Frank Smith headed for the doorway and then turned. "Bill? You better wrap it up here we are due at the dinner with Bob Mackey and his crew at 7:00 p.m. tonight."

Bill shuffled some papers on his desk, looked up at Frank, and responded, "I'll be there with bells on. This is a big move for us, and I wouldn't miss this dinner for all the tea in China."

"Great, Bill! I'll see you at the restaurant at seven sharp. I got to finish up a few things on my desk, and then I'm heading home to get ready." Frank opened the door and walked out of Bill's office. He stopped at Bill's secretary Sharon Smith's desk and asked her to remind Bill to pack it up no later than 5:00 p.m. Sharon acknowledged with a nod and a smile. "We'll see you there, Frank. He'll be out of here at five or I'll shoot him."

"Thank you, Sharon!" Frank replied as he walked down the hall to his office at a brisk clip. Frank sat down at his desk and called his secretary, Karen Mill, on the intercom. "Karen, would you please call my wife and confirm dinner arrangements tonight with her? Then call the dry cleaners and make sure my black suit is ready so I can pick it up on the way home."

Karen's voice replied through the intercom a second or two after she received his instructions. "Sir, your suit has been ready for pick up since 10:00 a.m. Mrs. Smith said she'd be ready by the time you get home. Is there anything else I can do for you, sir?" Karen was very efficient and left nothing to chance.

"No! Not right now, Karen. Thank you!"

Frank noticed some supply requisition forms on his desk that needed a signature but wasn't familiar with the project. He reached to push down the intercom button again to ask Karen what they were for. "Those requisition forms are for the Ahwatukee Hills site, sir! Mike needs those fittings delivered right away to complete the kitchen trim in those units," Karen reported as if anticipating his next request. Frank smiles and signed the forms and started to tell her they were ready.

Karen walked into the office and headed straight for his desk. She collected the signed requisition forms and headed out the door at a brisk pace. "Karen! Explain to me again why they need me in this office?" Frank commented as he observes her efficiency in the handling of the office.

"Because you're co-owner of this company, and people who work for you enjoy seeing you slaving away at the desk to keep them employed," she replied as though she had answered that question a hundred times for him.

Frank leaned back in his leather desk chair, put his hands over the back of his head, and retorted, "Oh, yeah! That's it."

Karen closed Frank's door and motioned to the office runner, Chip. "Chip! Run these forms over to Baker and have him get these pieces out to Mike's crew at Ahwatukee Hills. They need them right away." Karen turned to her intercom and pushed a direct line to Tom Baker in logistic support. "Tom? Chip is bringing over your requisition forms for the flange and fittings Mike needs at Ahwatukee Hills. Please ensure they're out to the site by 4:30 today. Thank you, Tom!"

Her top line rang once before she jumped on it. "Hello, Mike! The flange and fittings for the kitchen units will be on site by 4:30. You'll be able to finish those units tomorrow morning, and if I'm not mistaken, that gets you another schedule bonus and me a dinner. You're very welcome, Mike!" Karen nodded several times, smiling, no doubt receiving loud comments from Mike.

"That's right. I am the best, and you are very perceptive to notice. Mike, you and Mary are having dinner with Frank tomorrow at the Camelback Inn at 5:00 p.m. sharp. Yes! It's very important. I think you're getting a promotion, but don't tell anyone you heard that from me. I have a reputation to protect. Okay! Good luck!" Karen ended the call flopped down into her chair and threw her shoes across the room and let out a big sigh.

Bob, Mackey, and his entourage arrived at the Ritz Carlton in downtown Phoenix. They bailed out of their limousines like the secret service protecting the president. They scurried into the hotel where their envoy was waiting with an army of bellhops that rushed to retrieve the parties' belongings and hustled them into the elevator to the penthouse suite.

As they entered the penthouse suite, Bob Mackey shouted to Susan Cromwell, "Susan? Get everyone together in the blue buckle room by 5:00 p.m. sharp, dear. Then call the Rusty Pelican and have their wine steward pull out the best champagne he has and chill ten bottles."

Susan, with phone in hand, started contacting all the party members to standby for conference call. "Yes, sir! I'll take care of it."

Bob grabbed his black suit off the hanger trolley and headed into the bathroom. Susan followed placing his toiletry items on the counter in front of the mirror. Bob stripped down while the shower temperature was being adjusted. "Susan, please find my black tie with the red polka dots. I'm wearing that one tonight." Then he climbed into the shower.

"Yes, sir! I think I put that one in the blue case. I'll pull that out for you, sir. Will there be anything else?" she asked as she headed into the next room.

"No, thank you. That should be it for now. Oh yes! One more thing! Please wear that midnight blue dress of yours that I like so much."

Susan popped her head back in the bathroom door. "As usual, Bob, you have excellent taste. I'll wear the blue dress tonight." She left the room and walked down the hall to her end of the penthouse and started getting ready for the evening.

At the office, Bill Smith is finishing up some paperwork. The door swung open. Sharon stuck her head inside. "Bill, it's time to go. You have minutes before I promised Frank I would drag you out of here."

Bill took a final look at the papers on his desk then stood and grabbed his hat. "That will do for today. I'm out of here."

He moved toward the slightly open office door and opened it the rest of the way. "Sharon, are you ready, dear? We need to get a move on."

Sharon was standing by her desk with her sweater in her arm with a smile on her face. "Bill! When have you ever known me not to be ready to leave?"

Bill smiled and walked over to her and planted a wet one on her lips. "Silly me! What was I thinking?"

Bill and Sharon Smith headed out of the office arm in arm to the front of the building. They exited through the main entrance and walked straight to their car. Bill opened the car door for his lovely wife of twenty-five years and then walked around to the other side and climbed in. He started their '94 Mercedes 540 sedan up and drove slowly out of the parking lot.

The Smiths had lived in Arizona all their lives together. They went to the same high school and college and finally married after graduating ASU. They lived in Mesa for twenty years after they were married and moved to Gilbert just five years ago. Bill and Sharon pulled into the circular driveway of their four-thousand-square-foot home just south of Houston Road.

Their home was a beautiful Victorian-style home built by himself and his own crew. Behind their home, they had a clay tennis court, swimming pool, and a natural turf putting green. He had a team come in after the house was built and put in a full-size basketball court on the eastside of the home. It was a very comfortable home for them and their two boys.

Frank Smith pulled up to his home in Mesa just about the same time. His home had a modern, freestyle design with lots of windows. The neighbors teased them about living in a glass house. Frank Smith designed it himself and was very proud of the home. It was very energy efficient, and 65 percent of the home was constructed with double-paned, thermal-treated glass.

Frank pulled up to the front of the house and admired his work. Then he rushed in to find April Smith, his lovely wife, nearly ready. "Holy moly! I better a move on I can't leave a beautiful woman waiting!" He kissed her on the cheek and ran up the stairs to the huge master bedroom. He noticed that she had gone by the cleaners and picked up his black suit and had it laid out on the bed for him. "Shit! I knew I was forgetting something."

Frank always forgot to go by the cleaners. It just wasn't something he could program in his cluttered mind. The beauty of their twenty-year marriage together was that April had a mind like a steel trap. She never forgot anything. Frank had become so dependent on her that he joked about the cost to replace her. Frank jumped into the shower and got cleaned up for the big dinner.

Back at the Ritz Carlton, the CIG team was forming up in the Blue Buckle room. Coffee and snacks were laid out on the table along with bottled water and writing pads. Bob Mackey walked into the room and everyone stood. "Be seated, please. We have a few things we need to go over," Bob recited as he took his place at the head of the table. "Jeff? Do you have the changes to the schematics I asked you to bring?"

"Yes, sir! I have them right here." Jeff Corbel spread the blueprints over the table.

Bob examined them closely. "Yes! That's exactly what I want. The first building in our contract with Smith Brothers Construction is a split building. One half of the building on the West Side of Washington and the other on the east." Bob pointed to the two halves of the building on the blueprints.

Tom Porter looked at the blueprints and shook his head. "Bob? That's not what the contract we forwarded to them said. It clearly indicated that they would be responsible for the completion of one of the five units with an option to contract with us for the others if the first was completed by scheduled completion deadline." Tom points to the two building on the blueprint. "These are two distinctively separate units. It won't work under contract stipulations that we proposed to them."

Bob looked at Tom, nodding in agreement with him. "Yes, that's the way it looks. Except for one small detail—well, actually four fairly large details." Bob pointed to the lower left portion of the blueprints where a cut-away section was designed separately. "Right here, it clearly indicates that the east and west portions of the building will be attached together with four elevated glass escalators for access. These catwalks will be spaced approximately four to six floors apart and will allow unlimited access from one building to the next."

Tom read the notes in the corner and looked over the adjusted details. "Well, I'll be damned, Bob—you did it again. That was a brilliant maneuver. Who's going to present that to Bill and Frank Smith?" Tom asked while looking around the room guessing whom Bob as picked.

"Why, Jeff Corbel obviously—he's my operations chief. He'll do a splendid job presenting that tonight. What do you say, Jeff? You up to it tonight?"

Jeff Corbel was just about halfway through a cup of coffee when he heard his name mentioned. He gagged on the coffee and spit it out on the floor, carefully missing everyone in the room. The room broke out in light laughter. "I guess I'm okay with it, Bob! I would have liked a little more time to prepare, but what the hell!"

Bob patted Jeff on the back. "Atta boy! Okay, let's get down to the limousines and head over to the Rusty Pelican. I'm starving," Bob barked as he headed for the door. His entourage followed in single file with Jeff scooping up the blueprints. He was still choking a little bit on his coffee and wiping his mouth.

They filed into the elevator and headed to the garage where their limousines were waiting. As they arrived at the garage, they scattered into their respective limousines, and the carpool flew out of the garage. Ten minutes later, they arrived at the Rusty Pelican restaurant. The limousines dropped them off at the front entrance. Everyone waited together in front until the last limousine emptied its cargo.

They walked into the front entrance of the restaurant where the manager was waiting. "Good evening, Mr. Mackey. Please follow me. I hope you enjoy your stay with us this evening. Bill Smith and Frank Smith are already here." The manager escorted the party to their private dining room upstairs.

Bob Mackey and his entourage entered the plush upstairs dining area and shook hands with everyone seated at the dinner table. "Mr. Smith! Mr. Smith! It's nice to see that you both have the same taste in lovely women."

Bob smiled at their wives and kissed their hands before sitting at the table. "Thank you, Mr. Mackey! I can assure you, though, these two lovely ladies chose us, or we wouldn't have stood a chance!" Bill Smith shared, and the room broke into light laughter.

Jeff Corbel stood and placed the blueprints on a tripod in the corner of the dining area. "Good evening, Mr. and Mrs. Bill Smith." Jeff nodded to them then turned to the Smiths. "Mr. and Mrs. Frank Smith! And honored guests." Then he pointed to the blueprint for the first building to be constructed in New York as the new headquarters building for CIG.

"As you can see, we have a single building here that is divided into two full-sized structures and bridged from east to west with four glass catwalks that allow unlimited access to each wing of the building. The projected goal for completion of this project is the fifth of January 2002." There was no comment from the table, so Jeff continued, "Once the deadline is met on this contract, we will issue five additional contracts to the Smith Brothers Construction Company totaling 8.7 billion dollars."

Heavy applause filled the room after Jeff completes his presentation. "Are there any questions?" Jeff asked nervously.

"I have a question." Bill Smith stood at the table while the applause settled. "Can you get me 1.5 acres of storage and workspace for my crews separate from the site itself but close enough to prevent unnecessary logistic problems?" he stated, looking at the blueprints.

Jeff stuttered a little and glanced over at Bob Mackey. Bob nodded and turned toward Bill Smith. Then Jeff responded confidently, "That would be affirmative, sir." There was a quiet pause across the room, all eyes on Bill Smith.

"Then let's eat. I'm starving." And Bill sat down and pulled his napkin off the table and put it in his lap. The room erupted into applause, and Bill looked over at Frank Smith confidently and gave him a nod and wink. Frank smiled and nodded back.

Bob Mackey called for the champagne, and Jeff wrapped up the blueprint and handed it over to Frank Smith. The dinner continued, and the champagne flowed.

Next morning in the Smith Brothers Construction Company conference room, all the team leaders and construction crew bosses were sitting at the table. Bill and Frank walked in, and everyone stood and greeted them. "Be seated please. We have some business to discuss today that's very important to the future of this company," Bill shared as they took their seats at the table.

"Last night, we closed an 8.7-billion-dollar contract with CIG." They erupted into applause and cheers, everyone high-fiving each other and hooting their approval.

"Okay, please settle down. We have a few decisions to make on this project." Everyone settled back down into their seats, unable to remove the smiles from their faces.

"First, we need to ratify the appointment of our new project manager who will oversee the construction of all five buildings in New York. This assignment needs to go to a man who will be able to overcome logistical support issues and still meet the deadline. Frank Smith has selected Mike Evans. Is there anyone here that can see any reason this would not be the appropriate choice?" The room remained silent.

"Okay, Frank! You better tell him today so they can start packing for New York. People, I appreciate your support and wish you a fine day today. Let's get back to work." Bill stood, and Frank followed as they left the conference room together. Out in the hall, they high-fived each other, still excited about the new contract.

Bill stopped and turned to Frank in the hall. "Listen, Frank! Make sure Mike and Mary know that the doctors in New York are ten times better than anywhere else and we will cover all medical expenses for little Sarah." Bill handed Frank a packet from his jacket pocket containing information on a school for gifted children in the middle of New York City.

"I'll let them know that, Bill. Thank you for taking the time to get hold of this information on the school," Frank responded and turned to walk to his office.

Bob Mackey arrived at the airport at 11:00 a.m. sharp with a fleet of limousines. He stepped out of the car and was greeted by Bill Britt. "Good morning, sir! How was your meeting last night?"

"It went fabulously. Thanks for asking. We saved 2.3 billion on the contract bid for all five buildings. That means I can keep you around for a couple more days." Bob laughed as he walked toward the jet.

"Excellent, sir. I'll hold back my resume to Tucker Industries," Bill offered and started to laugh.

"Bill! You're a good man, and I love having you on the team. You don't need to send your resume out to anyone. You'll always have a position with me. That I can promise you. Take a few days off and relax and meet me in New York next week." Bob Mackey climbed the steps to the plane.

"Thank you, sir. I'll see you there on Wednesday," Bill responded as he backed up from the steps to the plane. Bob acknowledged his response with a wave just as he entered the plane.

Inside the aircraft, Bob asked the steward for a scotch on the rocks and sat down in one of the several leather recliners. Susan

Cromwell joined him and asked, "The pilot wants to know if you still want to fly over the Heber fire?"

Bob looked at his watch and responded, "Yes, I do! With that fire burning down the timber up there, I want to see if I can get some new bids on that property." Susan picked up the cockpit phone and acknowledged the route over Heber with the pilot.

The plane taxied out of the hanger and took off right away with no delays. Once airborne the pilot banks the plane sharp to the northeast toward Heber. Twenty minutes into the flight, the pilot buzzed back into the cabin. Susan picked up the phone and informed the pilot to put his instructions on the overhead intercom. The pilot announced over the intercom that they were flying over Heber, Arizona, and to look out the right side of the aircraft.

"Sad but beautiful! Jeff, I want to know what Smith Brothers Construction Company would bid on this project. Send them a fax and see what they can do. With all that timber burned out, their construction costs should be minimal now," Bob shared as he observed the destruction the fire had caused just outside of the town of Heber.

"Yes, sir. I'm on it!' Jeff responded as he started drafting a handwritten proposal.

Bob continued while staring out the window, "I want fifteen thousand acres to be converted into a luxurious, state-of-the-art clubhouse surrounded by homes and townhouses. We're going to need two lakes for boating and fishing and two eighteen-hole golf courses. We'll separate the event areas with natural timber from the area so it blends right into the surroundings."

Jeff continued to jot down the details of the proposal for the bid. "That is brilliant, sir. The public relations officer can present it to the Heber counsel as a facelift to the community." Bob instructed Susan to call his bank. "Susan, please call Brandon Sawyer and have him wire one million in aid to the citizens of Heber to rebuild the areas devastated by the fire and make sure that slips out to the media today."

The private jet banked to the north and climbed to cruising altitude for the flight back to New York.

Back in Phoenix at the Camelback Inn, Frank Smith and his wife were just arriving for dinner. They entered the plush restaurant and headed straight to the table reserved for them. Mike and Mary Evans were already seated and stood as their hosts arrived. "Good evening. Nice to see you two could make it on such short notice," Frank shared as he shook Mike's hand and nodded to Mary.

April Smith gave Mary a polite hug and kiss on the cheek. "How's little Sarah doing, Mary?"

"Oh, she is doing wonderfully. Thank you for asking." Both April and Mary took their seats at the table, and Frank and Mike followed suit. Mike asked Frank how the meeting went with the CIG.

"It went very well, Mike. Better than we could've hoped for. We landed an 8.7-billion-dollar deal. It's the largest contract we have ever had at Smith Brothers, and everyone is very excited about it."

"That's fantastic, Frank! I bet Bill is dancing a little jig right now," Mike replied as he scanned the menu.

"Well, to be quite honest, Mike, Bill's been dancing the jig since last night, and I don't think he's going to stop until next week." Both laughed as they mentally pictured Bill dressed up as a leprechaun at the St. Patrick's Day picnic last year.

"So I guess that means you have to choose a project manager for this deal? You know, I think Sam Macintosh would be perfect for this project," Mike offered modestly. Mary kicked him under the table, knowing that he was aware that the offer is going to him.

Frank looked at Mike seriously and then offered, "Mike! Bill and I want you to head up the project in New York. We need someone there that we can trust to handle the details and logistical challenges without calling home every five minutes."

Mike pushed away from the table and shook his head back and forth then stared at the kitchen door sternly as though his world was coming to an end. Frank's forehead beaded with sweat as he tried to calm Mike down. "We'll pay all your moving expenses and take care of all the medical and dental work for your family. It comes with a big raise, and it's your baby. Whatever you say goes." Frank waited for a reply.

Mike continued to stare at the kitchen doors of the restaurant as though waiting for something. Finally, the doors swung open, and the chef came out of the kitchen with a chocolate cake and candles burning on top. He was followed by several of the kitchen staff singing to the Burger King theme song.

"Have it your way, have it your way! Special projects don't upset us. Much more income and the bonus let us know that you love us. All we ask is that you let us have it our way. Have it your way, have it your way!" Their singing stopped as the chef reached the table and placed the chocolate cake down in front of Frank.

Mike, now sporting a huge smile, looked back at Frank and said, "Have it your way, Frank! Make a wish and blow out your candles."

Frank looked at the top of the cake and then smiled and blew out the candles. On the top of the chocolate cake written in white icing was the answer to his question. "Yes, Frank! I'd love to be project manager!"

"Well then, congratulations on your promotion! I think!" Frank shook Mike's hand and wiped the sweat from his forehead. Mary bounced in her seat and then planted a big, wet one on Mike.

April leaned over and kissed Frank on the cheek. "I think Mike had you pegged from the minute you walked in, dear." Frank laughed and leaned back in his chair, motioning to the waiter.

"Would you put this cake in a box for me? I want to put it on Bill's desk in the morning," Frank explained to the waiter when he arrived. "Bill will get a kick out of this one!" Frank looked over at the happy couple and pulled out the information on the school Bill wanted them to see. "Bill and I wanted Little Sarah to have the best school in New York, and we have prepaid her enrollment to St. Paul's School for gifted children. It's the finest school in the nation and helps children develop their artistic skills." Frank handed the information to them.

Mary ran around the table and gave Frank a big kiss on the cheek and hugged him. "Thank you, Frank! Mike and I have been talking about this school for the past year." Mary hugged April and whispered in her ear, "Thank you!" April just winked at her and gave her a big smile.

While the World Sleeps

While the rest of the world sleeps comforted in the knowledge that a joint unified task force is protecting them, comprised of a multinational strike force and intelligence gathering agencies from the US, Britain, Canada, Russia, Japan, Mexico, and other now free world non-combatant nations, a secret meeting was taking place at Naha, the capitol of Okinawa, Japan, located in the Ryukyus Island chain just south of Japan.

In the basement of the Ace-of-Clubs on Highway One, a group of terrorists, drug lords, and international organized crime families were meeting to discuss the next step in a series of attacks against the strongest nations. Those countries that support the international joint task force against drugs and terror were their primary targets.

They had been secretly gathering for days now, shuttled in by air and sea transports to unknown locations around the island, disguising themselves as local farmers and rendezvousing at the Ace-of-Clubs. In the basement, a large oval table was set up with packs of information for each guest to review as they arrive.

Chaired by the notorious Muhmad Sharif, leader and main financial supporter of the Tri-Qaida terrorist network with cells in every country, acting independently to cause mass chaos and confusion, around the table were the shadow leaders of the main drug-supplying nations of Cambodia, Afghanistan, and the Philippines, as well as those countries that were hosting the training camps of terrorists, primarily Syria, Libya, Iraq, and Iran.

Seated to his right was Archibald Gamboni, leader of the largest organized crime family in the world. He and his organization were responsible for the mass distribution of all illegal drugs in North and South America and all of Europe.

The leaders of the many violent gangs, supported and funded through the trafficking of illegal drugs and prostitution, were present and represented a significant danger to law enforcement agencies in the United States and other countries with a similar goal. They had significantly infiltrated every major city in the United States and operated independently under one leadership.

Leaders of the Mexican, Russian, Japanese, and Italian Mafia had chairs filled at the table. Cuba was also represented because it played a significant role in driving drugs into the United States and was also positioned to stage a significant assault against the United States. Cuba had supported the running of drugs into the United States and Canada for years.

Archibald opened the meeting. "Thank you all for joining us today on this momentous occasion. Within one year, we will have toppled the strongest government in the world. It has been a long, hard road, but we now have the tools and equipment to level the United States of America. Once they fall, the rest of the nations that blindly supported them will fall too. This is a day to celebrate!"

Everyone in the room stood and drank a toast to the upcoming collapse of the USA. "Seven thousand human bombs have been successfully positioned inside the towns and cities across America and on the eve of the pagan holiday called Christmas. There will be a series of explosions that will level every city standing on US soil, killing over 60 percent of the nation's population."

Cheers flooded the room as they raised their glasses again and toasted the downfall of the United States. "We owe a sincere debt of gratitude to Mr. Ortega of the international drug cartel for getting all the drug lords under one umbrella. Without his diligent efforts to provide a never-ending supply of the most devastating drugs to America's youth for the last twenty years, none of this would be possible."

Cheers echoed through the room, toasting Mr. Ortega. "Those kids are so fucked up they've no clue what's happening. The only thing they are interested in now is a free ride to happy land. Eighty percent of the youth in America is on drugs of one kind or another. They have no interest to take over now. Once they are removed, the youth of America will welcome us with open arms because we have the dope. God bless America!"

Muhmad Sharif stood and waited for the cheers to subside then offered, "I want to thank Mr. Gamboni and his families for their untiring efforts to strategically place his well-trained greed mongers into the CEO positions of the major US corporations. This was a brilliant move to undermine the ideal of true capitalism, teaching them the shell game and dumping millions of American families onto government subsistence.

"This move sorely taxed the government budget and prevented the necessary funding for a successful homeland security campaign. It also demoralized the American public, who worked most of their lives for a company only to lose everything they had acquired. Americans now have no faith in corporate America and are suspicious of every move they make. It was an absolute stroke of genius." Muhmad Sharif raised his glass to Mr. Gamboni and drank a toast to him.

"We also need to thank the man who has made the human bomb such an effective and reliable tool, Dr. Swanson." Muhmad Sharif raised his glass again toward a curtain in the back of the room. The shadow of a man was reflected on the curtain with his glass raised. "As our efforts to train and program walking, talking, and very mobile bombs continues, Dr. Swanson's identity must never be compromised to any of us or anyone else," Muhmad Sharif continued.

"Most of you should now be familiar with his most recent work at the US border town of Nogales, Arizona. He was able to program that young man in under an hour." Cheers arose through the room. "His impeccable programming allowed enough explosives through the border to complete the packages that been sent out and programmed by Dr. Swanson. These seven thousand packages will be in place and ready to detonate on the evening before Christmas day."

The room exploded with cheers and applause again. Muhmad Sharif smiled, tipped his glass, and sat down at the table. Archibald

Gamboni showed a slide of the United States on the wall. "As you can all see, the territories of the United States and Canada have been equally divided among all of you. I will assume the role of president and direct our new Congress and Senate to redefine the territories into seven separate states independently governed by you. You will all be kings in a new country."

The meeting continued well into the morning hours, as the final stages of the infamous plan were carefully set into motion.

CHAPTER 21

International Intelligence Meeting

John Pomeroy, director to the Central Intelligence agency; Peter Merritt, director of the Federal Bureau of Investigations; General Tom McLane, Director of the National Security Agency; and George Bell, director of Homeland Security for the United States, arrived at a secret location in Belgium traveling in an unmarked vehicle.

As they exited the vehicle, they hurried into a deli entrance. Hustling through the deli to the rear, they entered a storeroom and closed the door behind them. They were dropped three floors down into the ground to a secret conference room. As they entered the conference room that was guarded by United Nation forces, they took their seats at the table.

President Ferdinando Turmoni, director of the United Nations Security Task Force and Response Team, addressed the multinational intelligence-gathering service representatives. "Good afternoon, gentlemen! Thank you all for responding to the call so quickly. The reason for this emergency session of the Joint UN Security Task Force is to assess new intelligence recently acquired by reliable sources."

He turned to the screen and put up a slide of Archibald Gamboni, a notorious criminal mind who had evaded every attempt to be captured. He was known to have conspired to overthrow many of the governments represented in the room. "We are all familiar with this face and the name Archibald Gamboni. I'll turn the brief over to Dr. Reichstein, chief intelligence officer for the Blue Brigade, a highly effective stealth response team originally created for the protection of European interests."

"Thank you, President Turmoni. Gentlemen, our intelligence sources behind the lines in Afghanistan, Pakistan, and Somalia have discovered documents that indicate a significant strike against the United States, Canada, and Mexico within the next six months. This attack will be massive and very deadly to the populations of these countries. We expect that if the attack is allowed to be carried out it will result in a 60 percent fatality rate against those populations."

"The documents were uncovered during a Blue Brigade assault on suspected terrorists housed in the village of Utang in Somalia. Most of the Shiite terrorists killed themselves with cyanide capsules before they could be questioned. However, the documents stored in the homes were of such importance we felt it necessary to call this session."

CHAPTER 22

Authenticating the Thirteenth Scroll

Tim Wilford and Carl entered the Museum of Natural History after the Fire Chief determined it was safe to enter. It was about 3:00 a.m., and the rescue and firefighters had finally departed. Yellow tape was stretched across the entrance, and a police officer was guarding the front of the building.

Detective Grayson escorted Tim and Carl into the building and down the flight of stairs to the basement where the vault was. He was investigating the cause of the fire and insisted on staying with them while they were in the building. "I know you want to view the contents of the safe, and I have no problem with that. But you're not going down there alone. I'm going with you just in case there are clues as to why someone would want to blow your building up, Mr. Carlson."

"Fine then! Just please be very careful in the vault. There are very valuable documents in there that will fall apart if touched," Tim instructed the detective.

"I won't touch a thing!" Grayson responded.

As they opened the steel gate blocking the access to the stairs, Carl started to run his fingers up and down the steel flange on the gate door. "You got to be shitting me, Carl!" Tim barked, thinking that Tim had put another nickel in the gate door too.

"Yeah! I'm shitting you, Tim. I just wanted to see how much that annoyed you again." Carl unlocked the gate and pushed it open.

The three of them descended the wide stone staircase into the basement. It was pitch black without the power on. The only light was that provided by the three flashlights they are carrying in their hands. At the bottom of the staircase, Carl ran his light along the marble wall to the left of the stairs. He located a red button encased in a glass marked "For Emergency Use."

Carl broke the glass cover and pushed the button. The emergency battery backup lights illuminated the basement. In the center of the room was a huge vault. "There we go. That should help a little," Carl said as he walked straight over to the vault. He entered the combination to the safe, and the massive door released. It took all three of the men to push the door open since the back power wasn't connected to the electronic power-assist wrench for the vault door.

The light inside the safe revealed no damage to anything inside the vault. Not a single drop of water or ash made it into the safe. "I told you everything would be all right, Tim," Carl explained as a matter of fact.

"Yes, you did, Carl. Yes, you did!" Tim concurred as he glanced through the vault, trying to locate the shipment.

"It's over there, Tim." Carl pointed to the crate beside the vault door. "Thank you, Carl. Could you please give me a hand with this?" They lifted the crate up onto the table in the center of the vault. Tim took the crowbar from a hook on the wall and started to open the crate.

"Hello! Anyone down here?" a female voice echoed through the basement.

"Who the hell is that?" Tim asked, looking at Detective Grayson.

"I don't know. I told the blues to keep everyone out while we're down here," Grayson responded, looking at Tim with a puzzled look on his face.

"Well? Find out who it is. This shit is priceless!" Tim barked back at Detective Grayson.

"Okay! I'll check it out, but don't open that till I get back in here," Grayson demanded as he walked out to see whom the female voice belonged to. "Over here! Who are you, ma'am? More importantly, how did you get past my men upstairs?" Grayson questioned the female and checks her identification.

"I'm Dr. Jessica Parker, and I'm here at the request of Dr. Tim Wilford. Whom, I might add, left me stranded at the airport at 2:00 a.m.," she yelled loud enough so Tim could hear her in the vault.

"Oh, shit! I forgot about Dr. Parker's flight!" Tim exclaimed to Carl as they headed to the vault door together.

"Great, Tim. That's the way to make a lasting first impression on the person that's going to authenticate your life's work. That's fucking brilliant… I wish I'd thought of it."

"Hello, Dr. Parker. I'm so sorry I forgot about your flight in all the confusion here," Tim explained as he hurried to greet her. "This is Carl Curry, the museum curator, and I can see that you have already met Detective Grayson." With greetings exchanged, Grayson gave Jessica back her ID and shook her hand. "Ma'am!"

"We were just about to open the crate, Dr. Parker. Would you join us?" Carl asked and led the way back into the vault.

"What happened to the museum, Dr. Wilford?" Dr. Parker asked as she followed beside him.

"Well, we don't really know exactly, but it blew up. That's why Detective Grayson is here with us, and his men are outside," Tim briefly explained as they entered the safe, and he picked up the crowbar again.

"Right now, I'm more concerned about the scroll." Tim pried open the top of the crate and moved the packing material aside very carefully. There were thirteen containers in the crate; all of them appeared to be identical. Tim reached in and pulled out the thirteenth scroll. "Here it is, Dr. Parker. This is the scroll I told you about. If this is authentic, it's the written testimony of John the Baptist written in his own words."

"My God, Tim, this is a phenomenal find. Are you sure it's authentic?" She took the scroll case from Tim.

"I believe it is, but I need you to confirm it as fast as possible if this is the actual thirteenth scroll. There are a lot of evil people out there that will do anything to keep it from being discovered," Tim explained as he offered her the case.

Dr. Parker slowly felt the case material and smelled it. Then she carefully opened the case and removed the scroll in front of Carl,

Tim, and Detective Grayson. She had Carl hold the top end of the scroll as she unrolled it to see the writing. "Looks Greek to me!" Detective Grayson volunteered.

"Well, it's not Greek, Mr. Grayson. It looks like-old style Arabic." She moved her finger across the letters and paper, feeling the texture. Then she smelled the document, and her eyes lit up with excitement. "This is old papyrus, maybe two thousand years old. It has a very distinctive smell that is not reproducible even in the most sophisticated laboratory. I believe this is authentic, but I need to run some tests first. I need get this under better light to read it."

Tim and Grayson instinctively shined their flashlights on the papyrus for her. "Thank you! It says, 'I John called the Baptist by those who know me. Put to script my knowledge of my Lord, the man they call Jesus, the Messiah.'" Tears start to form in Dr. Parker's eyes as she started to read to herself for a couple of minutes.

Then she started to read aloud again. "That in the year of 2001, on the eve of our Lord's birth, a child will sing, and the world will stop. She will sing with Angels and every heart will open to the love of the Lord. Wars will end and peace will triumph over the land. The hungry will be fed, the naked will be clothed, and the homeless will be sheltered. No man will speak evil of another. The caged will be set free and hunger for the love of man. The lion will lie next to the lamb. The sunrise will bring forth new light onto the world such as is has never seen before."

Dr. Parker collapsed onto the floor crying. "What is it, Dr. Parker? Are you okay?" Tim reached down to comfort her.

"Yes! I'm fine. Didn't you hear what I read to you?" she responded angrily.

"Yes! I heard what you read. But why are you crying?" Tim asked, still puzzled.

"You really don't understand, do you? None of you understand what this is saying." They looked at each other, trying to reflect on it to see if they missed something.

"No! I guess we don't really know what it's saying."

Dr. Parker stood up and looked over the scroll again, trying to locate where she left off. Then she started to read again. "In this time, I shall give to the world a child. Deaf and dumb to the world a message of love with all the Angels in heaven. Every ear shall hear the voice of 'SAVE' the little angel," she continued, trying to pick up where she left off.

"This will be my sign to the world to have hope for the time I shall come back is near, very near to them. Rejoice! I say rejoice and offer up song and praise to the Lord God Almighty for he so loves the world." She stopped and scanned through the text of the scroll, and then she continued, "This is the first of three signs to establish the path of the Lord Jesus the Christ that will be presented to the whole world. Again, rejoice for the Second Coming of the Lord is near."

Dr. Parker finally finished scanning the document, and then she tasted the corner. "This document is authentic—I have no doubt—and I believe the chemical tests will only confirm my certainty of this."

Tim jumped up and screamed, "Eureka! That is fantastic! I knew all along that it was the real McCoy. I can't wait to call Haseeb and tell him." Tim hugged Dr. Parker and his friend Carl. Then he started to hug Grayson, but he backed away and put his hand out for him to shake.

"We don't hug guys in my family. We shake hands." Tim shook his hand then pulled him in tight for a big bear hug. Detective Grayson was caught off guard and didn't know whether to smile or shoot.

"Dr. Parker, how soon do you think you can complete your authentication on this document? It's very important that we release this discovery as soon as possible," Tim asked Jessica with a degree of urgency.

"I can probably overnight a sample to the lab to get the carbon dating done in a week. In the meantime, I can do a chemical analysis here if you have the right equipment."

Tim turned to Carl and waited for his response. "I do have a small laboratory on the other end of the basement used for chemical testing, but I'm not sure that we have all the chemicals that you would require," Carl responded, shrugging.

"That will do perfectly!" Tim responded quickly. "I can get you any chemicals that he doesn't have here in a couple of hours."

Dr. Jessica Parker nodded and took a small piece of the parchment from the corner of the scroll and put it into a plastic sandwich bag she retrieved from a box on the vault door. "Well, that's all I need here." Then she turned and exited the vault followed by Detective Grayson. Tim and Carl rolled the scroll up carefully and placed it back in its leather pouch and placed it back in the crate. Tim then sealed the crate back up using the crowbar as a hammer.

Tim and Carl exited the vault and secured the vault door behind them. "Don't let anyone open this vault without me being present, Carl," Tim insisted as they joined the others.

"You got it, Tim." Detective Grayson turned and looked back at the vault. "Is everything okay in the vault? We only took a look at that scroll in there."

"Yes, it's all good, Grayson, trust me," Carl replied as they started toward the opposite end of the basement looking for the laboratory.

"Carl, can I borrow your cell phone for a moment? I need to call Haseeb," Tim asked quietly, almost whispering.

"Sure, here, just punch out this phone number plus the overseas operator, and she will connect you. This is going on your tab, Tim, 'cause it's going to cost a fortune," Carl explained as he pulled his cell phone from his belt and handed it to him.

Suddenly, the lights went out in the basement, and the flashlights failed too. The air filled with the horrible stench of rotting flesh. It was pitch black; they couldn't see their hands in front of their face. "Everyone stay where you are and don't move!" Detective Grayson shouted as he reached for his weapon. But before he can put his hand on it, he was struck from behind and fell to the floor unconscious.

"What the fuck is that disgusting odor?" Carl yelled out as he moved toward Grayson's position.

"Oh, shit!" Carl blurted out as he fell to the ground. Dr. Parker screamed frantically as if that would scare off the bad guy.

"Jessica! Pull yourself together. I can't hear anything but your infernal screaming," Tim yelled out in her direction. Dr. Parker's scream subsided to a soft whimpering. "That's much better. Carl!

Are you okay? Where are you?" Tim put his hands out toward his last position.

"I'm down here on the floor with Grayson. He's out cold." Carl stood up then whispered, "Tim, where are you?"

"I'm right here in front of you. Just reach out and grab my hand, and we'll walk over to where Dr. Parker is." Carl reached his arms out and grabbed on to Tim's hand. "There you are, buddy. Am I glad I found you. You know, you really need to get a breath mint! Your breath could knock a buzzard off a shit wagon!"

Tim was still reaching out for Carl and heard him talking to him. "Carl? I'm over here, asshole. Who are you talking to?"

Carl screamed out and then hit the floor with a thud. "Fuck! What the hell is going on here?" A deafening roar filled the room, and the stench of foul odor thickened.

"My God! That stench is rancid. Who the fuck are you and what do you want?" Tim screamed out, having nothing better to do.

Jessica shuffled toward the sound of Tim's voice, still whimpering. "Oh, thank God, Tim! I thought you were the monster."

Tim turned toward the sound of Jessica's voice. "Jessica? I'm over here." Then she screamed, and another piercing roar echoed through the basement and the sound of Jessica hitting the solid concrete floor. "You fucking asshole, stand up and fight like a man, you pussy," Tim yelled out as though he alone could conquer the beast.

The lights flickered a few times and then came back on. Tim was wishing now that they hadn't come on. Standing in front of him, a few feet away, was the most hideous beast he had ever seen. With gruesome snot dripping, the demon from hell was standing over the bodies of his friends. Tim was not willing to give up as easily as his friends did. Tim had seen Indiana Jones fifteen times and knew that he couldn't show his fear.

"So what the fuck do you want, shit breath? Dental floss?" Tim yelled out as loud as he can, attempting to put the fear of God into the demon from hell. The creature seemed to be responding to his challenge. It placed its dripping hand over its mouth and blew into it then sniffed his breath. The creature fell backward a few steps and almost lost its balance.

It shook off the dizziness and let out a bloodcurdling scream and stepped forward toward Tim. "Then we agree. You definitely… need some serious dental work. Do you have a good dental plan where you work?" Tim conversed with the monster, trying to buy enough time to shoot up the stairs and out of the building.

The creature cocked his head to one side, trying to figure out why Tim wasn't terrified of him. Then he dropped down low and crouched, and in the blink of an eye, he sprung into the air straight toward Tim. Tim only had a moment to swallow the large lump in his throat. He realized that he would never see the light of day again. Then defiantly, Tim flipped the monster the middle finger. "Fuck you, asshole!"

The creature exploded into a thousand black beads and circled into a small vortex. Then it was sucked into the concrete floor and disappeared. Tim looks at his finger with his mouth wide open, thinking that he had shot the beast with an invisible gun.

"Ahem!" a voice rang out behind him. Tim turned to see the image of a large Navajo Indian standing behind him with a bow and a quiver of arrows.

"You got to be shitting me!" Tim said as he looked at the Indian all decked out in original Apache warrior attire.

"No shit! Me Geronimo! Chief of the Chiricahua Apache nation."

Tim was still not sure if he was awake or dead as he looked him over. "What are you doing here?"

Geronimo slung his bow over his shoulder. "Here to help you!"

"Okay! Why do I need help? What exactly is going on here? Am I dead?" Tim asked, becoming more confused with each passing moment.

"You are not dead. Me here to help you," Geronimo answered in half-sentences.

"Listen, Geronimo or whoever you really are, Indians don't talk like that anymore. They speak English just like me. Now please! What was that snot rag doing chasing me, and why did it kill my friends?"

Geronimo explained why he was here and tried to answer Tim's questions. "Listen, pal, I'm here to cover your ass. A great shaman called me back from the spirit world to keep you alive. Your friends aren't dead. Not yet anyway. And the creature that attacked you guys comes from that place of fire and brimstone you like to call hell."

"All right, now we're making some sense. So why does the devil want me dead?" Tim asked boldly.

"Because you discovered the thirteenth scroll, and he didn't want anyone to find it. If it is released to the people that God loves, then he will show his love with a miracle soon. People will start going back to church and worshiping God like they should have been doing all along. It'll kill business for Lucifer, and well, quite frankly, that just pisses him off."

"So it is real? And the miracle is for real too?" Tim inquired.

"Yep!" Geronimo replied, not wanting to draw this out.

Tim pulled a chair over and sat down, trying to think of more questions. "If Lucifer wants me dead, why hasn't he killed me yet? He has a pretty big army of those ugly monsters, doesn't he?"

Geronimo sat down, realizing that this could take all night. "Yes, he does have an army of creatures that can do unspeakable things to you. But he must play by the rules. He can't just kill the keeper of the scroll and take it from them. They must give it to him willingly in exchange for something, usually their life. Then once he has possession of it, he can destroy it or, even worse, he could alter it." Geronimo put his head on his knees, waiting for Tim's next question.

"Okay! I kind of got the picture now, except one thing doesn't make any sense! Why me?" Tim asked, surprised that he was even having this conversation.

"Because you have a pure heart and soul. The keeper of the thirteenth scroll—your term not mine—must have a pure heart before he can touch it. Otherwise, you would be dead." Geronimo looked at his fingernails, waiting for the next question.

Tim thought for a minute then said, "What happens after I release the discovery to the news media, and everyone knows the contents of the scroll?"

Geronimo scratches his head for a moment then responds. "That's a good question, Tim. I don't know. But I'm sure of one thing: You need to get that out to the world as soon as possible. I got your back no matter what happens."

Tim smiled then reached out to shake the hand of his guardian. "Thank you, Geronimo. That means a lot to me." Geronimo reached out and shook Tim's hand. Tim felt a soft brush against the palm of his hand followed by a tingling sensation. "Wow! That's different."

Geronimo smiled and said, "You got that right, Tim. Listen, if you don't have any more questions, I've got a hot date with Hiawatha over on cloud nine."

"Oh, sure…sorry…go ahead. I think I can manage here. But what about my friends?" Tim asked as Geronimo started to fade away.

"They'll be fine. Have them take two aspirin and call me in the morning." And *poof,* he disappeared. Tim ran over to where Jessica was lying on the floor and lifted her head up.

A few moments passed, and they started to come around, moaning and holding their heads. "What the hell happened?" Carl asked as he tried to stand and fell back on his butt.

"You wouldn't believe me if I told you, Carl," Tim replied as Jessica started to sit up beside him.

Carl, Jessica, and Tim huddled together on the floor, trying to make sense of what just happened to them as they watched Detective Grayson come to. Grayson jumped to his feet as soon as his eyes opened and finished pulling his weapon out of his holster. He became very dizzy and fell backward onto the floor again as a shot rang out, and he hit the floor. The fall knocked him out for the second time.

"We really don't need to hang around for him to come back, do we?" Carl asked Tim.

"No, we'll hear him when he gets up again. Let's go find the laboratory and see if you got all the chemicals we need," Tim replied as he helped Dr. Parker off the floor.

CHAPTER 23

The Flight

Six-year-old Sarah Ann Victoria Evans was helping her mother, Mary, pack up her room. Sarah put her favorite doll on the floor in front of her, and using both hands, she signed to her mother, "Can I take twinkles in the car?" Mary smiled and nodded affirmatively to Sarah. They completed the last box in her room and walked into the hallway together. Mary turned to Sarah and looked directly at her. "Okay! I think we are done! Do you want some ice cream, Sarah?"

Sarah jumped up and down with childlike enthusiasm and raced her mother to the kitchen, dodging the boxes lined along the wall. Sarah reached the kitchen first and climbed into her seat at the table. She set Twinkles down next to her and placed her chin in the palm of her hands.

Mary arrived and walked to the refrigerator and opened the freezer. She pulled out the vanilla ice cream and held it up, looking at Sarah. Sarah shook her head. Mary put it back, pulled out the chocolate ice cream, and received an overwhelming vote of support. She dished up two heaping bowls of ice cream, set them on the table, sat down next to Sarah, and bowed her head for grace.

Sarah mimicked her mother and put her hands together in front of her. Mary said grace, and they dived into the ice cream together, making funny faces as they chowed down together. Mary admired how well-behaved young Sarah was. She admired how big she was getting now. It seemed it was only yesterday that she was holding little Sarah in her arms at the Desert Samaritan Hospital surrounded by reporters.

Sarah and Mary were very excited about going to New York. Mary knew Mike would be home shortly, and they would turn the packing over to the movers and head to the airport to catch their flight. After finishing their ice cream, Mary and Sarah held hands and walked through the house together, making sure that all was in order.

After walking through the house, they ran upstairs to the bathroom and started cleaning up and getting dressed. Sarah and Mary did everything together; they were inseparable. As they stood in the bathroom together in front of the big mirror brushing each other's hair, Sarah signed to her mother. "Mother, I'm excited about going to New York and meeting new friends."

Mary smiled and nodded her excitement too. Then Sarah added, "I'm also very happy to be going to the new school Uncle Bill and Uncle Frank put me in."

Mary looked at Sarah and supported her excitement. "Sarah, we are going to have a wonderful time in New York, and you will meet a lot of new friends there. Your new school has music and art classes as well as special writing courses that will help you communicate with everyone."

Sarah smiled and hugged her mother. Then she signed, "I love you, Mother!"

Mary gave her a big kiss and a second hug. "I love you too, sweetheart."

The front door slammed, and Mike yelled out to Mary and Sarah, "I'm home! Are you kids ready to go to Disneyland?"

Mary looked at Sarah and said, "Your daddy is home!"

Sarah ran out of the bathroom like a rocket and sailed down the stairs toward the front door. As she reached the bottom steps, she jumped up toward her dad. He caught her in his arms and gave her a big kiss and hug.

Then he put her down and said to her slowly so she could read his lips, "Where is your mommy?" Sarah grabbed his hand and dragged him up the stairs into the master bathroom. Mary and Mike hugged and kissed each other like adults did. Mike groped her buns of steel, as he liked to refer to them. Sarah played; she put her head down, pretending not to see him. Then he looked out of the corner of his eye and noticed her little game and put Mary in the shower and pulled the curtain around them.

Sarah laughed at them and danced up and down, still excited about the upcoming plane trip. She'd never been on a plane before and was very excited about the adventure. Mike released Mary and climbed out of the shower and grabbed Sarah. He gave her a big kiss and a hug and took her into the shower, pulling the curtain around them just like he did with Mommy. Sarah opened her mouth and laughed like only she could. It was not really a laugh; she just mimicked the motions and grunts. But Mike and Mary knew it was her way of laughing and loved to hear it.

Juanita Hernandez opened the door to the master bathroom after hearing all the commotion and entered. "Hello, Mr. Evans. Everything is packed for you and your family, and the car will be here in twenty minutes." Juanita Hernandez was their live-in aid for Sarah and the Evans family provided by Smith and Smith Construction. Bill and Frank thought that they could use the help around the house and someone they could trust to take care of Sarah when they went out.

Juanita was very sad to see them leave 'cause her family was here in Arizona and wouldn't be making the trip with them. Juanita's eyes were red from crying and packing most of the day. She was very sad to see the family leave. She was happy, though, for Sarah and her new school. She believed it would help her with her many talents.

"Juanita? Are you still crying? Please don't do that. We're going to be fine, and we'll miss you very much, but we will write you and let you know how we are doing. When you have some free time, you're more than welcome to come to New York and visit us especially around the holidays," Mary offered as she walked over and hugged her. They both started bawling together.

Mike looked down at Sarah, who was becoming a little emotional too. He thought to himself that women should not be allowed to hug at moments like this. It created a discharge from them that took hours to mop up. Then he walked over to the two of them and gave them a group bear hug and growled. Sarah joined in but could only squeak, which brought laughter into the mix.

"Come on, ladies. We have to get going. The car will be here any moment," Mike shared as he headed out of the room and down the stairs to the front of the house where the suitcases were. Mary followed close behind, and she bounced down the steps like a schoolgirl.

Sarah looked up at Juanita with a sad face then turned and walked over to the window, looking out over the neighborhood. Juanita sadly observed her from the doorway, knowing she needed a moment to herself. Sarah investigated the big tree next to the window and saw a little bird in the nest. The little sparrow noticed Sarah and looked at her with curiosity then jumped off the nest and flew toward her.

The little sparrow hit the glass with a thump and fell to the ground below, dead. Sarah jumped back, a little shocked at what happened, and then looked down at the bird lying in the grass. Juanita walked over to comfort Sarah after observing the incident. She put her hands on Sarah's shoulders and gave them a gentle, comforting squeeze to let her know she was there.

Sarah turned to see Juanita observing the dead bird on the ground with a tear running down her cheek. Sarah pointed her finger at the dead bird and moved it in a circular motion as Juanita watched with concern. Suddenly, the dead bird came to life and flew back up into the nest in the tree, where it was before.

Juanita made the sign of the cross and recited, "Jesus! Mary! Joseph! Pray for us!" Then she kneeled beside Sarah and looked into her eyes, shocked at what she saw. Sarah turned and met her gaze then put her finger to her mouth to keep Juanita silent. She smiled at Juanita and gave her a big hug and kiss then turned and walked out of the room. Juanita was frozen, trying to digest what she had just observed. She stood and crossed herself again and walked out of the room to follow Sarah down the stairs.

Mike had the bags loaded into the limousine and said his goodbyes to Juanita and opened the door for Mary and Sarah. Mary and Sarah said goodbye to Juanita and entered the limousine. Mike offered a final wave and followed them. The limousine started down the driveway, and Juanita watched it slowly drive off down the driveway. Sarah popped her head up in the back window and made eye contact with Juanita.

Juanita was uncomfortable at first after she witnessed it, then Sarah formed the words "I love you" and waved goodbye. Juanita mimicked the gesture and smiled as she waved goodbye to Sarah. The limousine turned the corner at the end of the driveway and drove off toward the airport. Juanita started to walk back into the house, thinking she had imagined the entire incident. Just then, a small sparrow landed on her shoulder and started singing, reminding her that what she saw was real. "Oh, sweet Jesus! That child is something special for sure!" she mumbled as she looked at the little sparrow still sitting on her shoulder.

The limousine arrived at the airport, and the porter grabbed their bags at the curb and tagged them for their flight to New York. Mary took a small carry-on bag and headed to the main door with Sarah. Mike tipped the porter and followed. They already had their tickets and headed straight to the gate their plane was leaving from.

The public address system announced their plane would be leaving for New York in thirty minutes. Mike, Mary, and Sarah walked together, holding hands through the security checkpoint and straight to their gate. Mike walked up to the boarding ticket counter and grabbed their boarding pass. They were traveling first class straight through to New York.

Sarah was a beautiful little girl. She had shoulder-length blond hair, sky-blue eyes, and a wonderfully petite little frame. Anyone who gazed on her instantly melted to butter. Sarah had no idea of the effect she had on people. She was very excited about flying in her first airplane and ran to the big window to see all the jets.

The call went out for all passengers holding tickets to board the airplane now. Mike stood near the gangway, waiting for Mary and Sarah to catch up to him. Sarah's head was everywhere, looking at all the new sights. As they entered the plane, Sarah stopped at the door and stared in at the huge entrance. For her, it was like walking into a castle for the first time. She took her time entering and walked down the aisle to her seat.

Mike, Mary, and Sarah were all seated together in the third row of the first-class section. The seat was leather and almost as big as Sarah's bed. She sat between Mom and Dad in the center seat. Unable to sit still while Mike put the carry-on in the overhead compartment, she jumped in the seat, looking at everyone behind her.

Mary helped Sarah sit down and fastened her seatbelt. Then she fastened her and turned to Sarah and patted her on the leg. "Here we go to our new home." Sarah smiled and leaned against her mother to get comfortable for the six-hour flight. Mike sat down and fastened his seatbelt and then double-checked Mary and Sarah.

"Are you ready for this, honey?" Mike asked Mary.

"Oh yeah! I'm ready. I feel like a little kid again," Mary replied as she leaned over and kissed Mike.

"Good, honey! I'm glad you're happy. I think Sarah is excited too," Mike said as he patted her on the head and messed up her hair.

Finally, the plane doors were closed, and the air started to blow through the open vents overhead. The plane rolled into position and began to taxi down the runway. Sarah's eyes lit up like she was seeing a Christmas tree for the first time. The loud roar of the engine signified they were about to take off. Then the plane started down the runway at full speed. After a few bumps and shakes, the plane glided into the air, and they were off to New York.

Three hours into the flight, Mary and Mike were snoring like lumberjacks. Sarah sat between them, coloring in a color book the stewardess had given her. There was a commotion in the back of the plane in the coach section. The stewardess from first class picked up the microphone and radioed the pilot. The co-pilot exited the cockpit and walked to where the stewardess was.

"What's wrong, Connie?" he asked quietly with Sarah looking right at him.

"There's a gentleman in the back complaining of chest pains. Wanda is with him right now. Should I get the paddles from the kitchen?" she asked as she walked down the aisle with the co-pilot.

"Did you check the passenger manifest to see if we have a doctor on board or a nurse?" the co-pilot, Tom McKann, asked as they headed toward the passenger.

"Yes sir! We have no medical personnel on the manifest," Connie Corkle responded as she led the way.

"Okay! You better grab the paddles just in case," Tom explained as he moved past her and headed to where the victim was seated.

Sarah could sense something is terribly wrong and slid past Dad into the aisle. She walked back toward the rear of the aircraft where the commotion was. Several people had surrounded an elderly man in his seat. He appeared to be having trouble breathing. His wife, sitting next to him against the window, was frantically trying to help him loosen his tie and shirt.

The co-pilot took a knee next to the gentleman and asked everyone to remain calm. "Please, please! Everyone remain in your seats and clear the aisle. The stewardess is bringing back the paddles, and we'll have the plane on the ground soon." He turned his attention to the elderly man, who was very pale in color. "Sir? Can you hear me? Sir?" Tom asked as he checks the man for a pulse on his neck.

His wife told the co-pilot what happened. "His name is Joe Morgan. He's thirty-nine years old, and he's never done this before. We were just sitting here talking, and he started having pains in his chest and trouble breathing." Then Silvia Morgan, his wife, began to cry.

"Ma'am, please. Everything is going to be fine. He still has a pulse, and he's getting some air."

The co-pilot pulled down the oxygen mask from the overhead and put it over Joe's face. Then he looked back to see where the stewardess was with the paddles. "Please, ma'am, just keep this mask on his face, and I'll be right back." The co-pilot hustled back to the front of the cabin, trying to locate the stewardess with the paddles. He passed right by Sarah as he heads to the front of the plane.

Sarah walked back to where the man was gasping for breath and saw his wife with tears rolling down her cheek. Sarah placed one hand on his chest and the other on his wife's hand, which was holding the oxygen mask. "Honey! You shouldn't be here to see this! Someone, please show this little girl back to her seat!" Silvia said with great effort, trying to hold back her emotions in front of Sarah.

Sarah smiled at Silvia and looked at Joe's eyes and continued to keep her hands in place. Silvia Morgan was surprised to feel her hand tingling then looked at Sarah. "What are you doing, honey?" she asked in a calmer voice. Sarah just smiled and continued to hold her hands in place. After a moment, people watching started to get a little nervous and moved toward Sarah.

Silvia Morgan shouted to them, "No! Leave her alone. Something is happening! Let her be!" Sarah looked at Silvia with a warm and comforting smile then leaned forward as though whispering something to Joe. She lets go of Silvia's hand and Joe's chest and stood back a step. Joe Morgan heaved up and down in his seat three times violently and then stopped gasping for air.

Everyone was amazed to see Joe no longer reeling in pain. He slowly opened his eyes. The color was back in his cheeks, and he was breathing normally again. Those that witnessed the miracle started to clap and cheer. Joe reached over to Sarah and gave her a big hug. "Thank you. You saved my life! But how did you do that?"

Sarah did not want them to know that she couldn't talk, so she kissed the man on the cheek and went back to her seat. The co-pilot ran past her with the paddles in hand. He approached the man, thinking he would have to perform emergency medical procedures to keep him alive. Then as he got closer, he noticed the man was fine and acting as nothing had happened.

"Sir, are you okay?" the co-pilot inquired as he arrived at his seat.

"I'm fit as a fiddle, sir! Thank you for your concern, but everything is fine now," Joe replied as he leaned over and hugged his wife. She was very happy to have him back and wasn't going to make a fuss about this if the little girl didn't want her to. She just knew Sarah didn't want anyone to know what she had done. When Sarah touched her, she heard Sarah tell her name and ask her to keep this to herself.

"When I left, you were near death, sir. What happened that changed that?" the co-pilot demanded to know. "I couldn't tell you, sir. When you left, I thought I was dead, but that little girl touched me, and now I'm fine," Joe Morgan tried to explain to Tom McKann.

Joe could see on Tom's face that he didn't believe him and wanted an answer that would make sense.

"Honestly, sir, I have no idea what she did. She touched me and whispered in my ear that everything would be fine, and it was," Joe continued, trying to add to his explanation.

"What girl? Where is she?" Tom looked around to see whom they were talking about. Everyone seemed to know what happened but couldn't tell the co-pilot who she was or where she came from.

The co-pilot carried the paddles back to the compartment where he got them and advised the stewardess. "All is well back there now! They said a little girl touched him, and he's fine. Did you see a little girl back there?" Tom asked Connie, still puzzled about the incident.

"No, sir! I was looking for the paddles. They weren't in the compartment that they were supposed to be in."

"Well! That's the strangest damn thing I've ever seen in my entire life. One minute the man was near death, no hope of medical aid in time to save his life, and the next thing, he's hugging his wife and enjoying the flight. What am I going to tell the captain?" he shared as he looked to Connie for a good story.

"Don't look at me! I couldn't come close to that one. You're on your own with this. Sorry!" She patted him on the shoulder and pushed him gently toward the cockpit. She took the paddles from him and explained she'd put them close by just in case. He nodded and entered the cockpit. "You'll never guess what happened on my way to coach today, Captain!" he announced as he closed the cockpit door behind him.

CHAPTER 24

Welcome to Manhattan

Mary, Mike, and Sarah arrived at New York's JFK airport right on time. They exited the plane and headed straight to the baggage claim area. As they entered the terminal, they saw a man in a black suit holding up a large white sign with their name on the board. Mike walked up to the gentleman and stated, "I'm Mike Evans! Who are you?"

The tall man in the black suit put the sign down to his side and shook Mr. Evan's hand. "Hello, sir. I'm Juan Ramirez with CIG. Mr. Mackey asked me to pick you and your family up and drive you to the apartment in Manhattan. It's just an hour from here, and we can leave whenever you and your family are ready."

"Thank you. That is very nice of him. We have some luggage to grab, and we should be ready to leave after that," Mike replied as he turned and introduced his family to the driver. "Mr. Ramirez, this is my wife, Mary, and my daughter, Sarah."

Juan shook their hand and smiled. "It's a pleasure to meet all of you. Please follow me. I'll take you over to the baggage claim area and we can get your bags."

Mr. Ramirez led them down the wide corridors of the airport crowded with passengers rushing here and there. After ten minutes of walking, they arrived at the baggage claim area, where the luggage was spinning around the oval belt. "There they are over there!" Mike pointed out his luggage on the belt. Juan grabbed a free dolly, loaded the luggage onto it, and led them out of the airport to a waiting limousine.

The limousine driver opened the doors for Mike and his family as they arrived. "Good evening, Mr. and Mrs. Evans. Did you have a nice flight?" the driver asked, helping them into the limousine.

"Yes! We did, thank you!" Mary replied as she helped Sarah in the car. The driver continued to hold the door open till everyone was inside, including Mr. Ramirez, and then shut the door behind him.

The driver walked around and got into the driver's side, and the limousine pulled away from the curb, heading toward Manhattan. Inside the limousine, Mike expressed his appreciation to Mr. Ramirez for greeting his flight. "Thank you for meeting us. We were a little intimidated by the size of this airport."

"My pleasure, sir. It's very nice to meet the man that is going to help construct our new office complex," Juan replied, and he looked out the window at the city lights in the distance.

"Well, I didn't expect to be greeted with a limousine to be honest with you. This is a nice touch, and I want you to let Mr. Mackey know that I appreciate it," Mike replied as he examined the inside of the limousine with all its fancy features.

"Mr. Mackey wants you and your family to feel right at home here in New York. He told me to express his appreciation to you for accepting the assignment. He has seen your work and is very impressed," Juan explained as he turned back and looked at Mike and his family.

"Honestly, I'm the one who is impressed to have such an opportunity. Between you and me, this is the biggest project I've ever been given, and I'm very excited to have it," Mike shared with Juan.

Juan looked over at Sarah, who had remained silent through the whole encounter. "Little Miss Evans, are you all right?" Juan asked politely. Sarah saw his lips and nodded affirmatively with a big smile.

"Sarah is deaf, Mr. Ramirez. She can't hear you, but she reads lips very well," Mary volunteered as she fussed with Sarah's hair.

"I'm sorry, Mr. Evans. I wasn't aware of that."

"No reason to be sorry, Mr. Ramirez. She doesn't really mind it at all," Mary responded, feeling she made Juan uncomfortable.

"Please, I would appreciate it if you all were to address me by my first name, Juan." Juan smiled at Mary and looked back at Sarah.

"My son Hector is deaf too. He is seven and goes to the same school that you will be attending, Sarah." Sarah smiled again at Juan and started to sign a reply.

Mary started to interpret her response to him. "Sarah says Hector is a nice name and would meet him."

Juan turned his attention back to Mary and smiled at her. He signed to Mary, "I have been signing for six years with my son, Mrs. Evans."

Mary laughed and grabbed Juan's hand. "I'm so sorry, Juan. I'm just so used to being a mouthpiece for Sarah that it just slips out," Mary said apologetically.

"Actually, there is no need for an apology, Mrs. Evans. I do the same thing with my son Hector all the time. Believe me, I understand perfectly." He smiled and released Mary's hand.

"Please call me Mary! I'm not big on titles, and I don't feel comfortable being called Mrs. Evans." She sat back in the seat and relaxed a little.

"Please call me Mike as well. I'm not excited about being addressed as sir or Mr."

Juan nodded and sat back in the fine leather seat. The car continued into the city, and the lights of the city cast a dome of light over it. It looked like the city was protected from the elements by a force field of light.

Finally, the limousine pulled onto Michigan Avenue in Manhattan and rolls up in front of a duplex with the front lights on. The front of the house looked like an old Victorian-style home squashed to fit the size of the street. A series of steps led up to the doublewide entrance to their new home.

Mike got out of the car first and stood on the sidewalk, admiring their new lodgings. "It's very impressive and stylish for a duplex," he offered as he crossed his arms and stared at the entrance to the home.

Mary got out and joined him then put her arm inside his. "It's lovely and quaint." Sarah jumped out and ran up the front steps to the door.

Juan handed the house keys to Mike and Mary. "My wife, Amanda, and I took the liberty of filling the kitchen with some food items for you. We arranged the furniture and put away the items that you had forwarded. This was so you and your family won't have to worry about unpacking right now. Just enjoy your new home for few days and call if you need anything." Juan handed Mike his card, and on the back of the card was his home phone number.

"I don't know what to say, Juan. You have been an enormous help to me and my family," Mike replied, surprised at the length that Juan went to make them comfortable.

"Just say yes!" Juan says with a big smile.

"Okay, yes!" Mike said, not really understanding why.

"Thank you, Mike. You have made me very happy. My wife is going to take me dancing now," Juan replied with great excitement.

"Why, what have I agreed to? Is it legal?" Mike said, showing apprehension about agreeing so quickly.

"To let me be your crew chief. I promise you, it's the best decision you've made today," Juan reassured with complete confidence.

"Excellent! That's one detail I don't have to worry about anymore," Mike replied confident he made the right decision.

"You and your lovely family have a good evening and settle in. I'll pick you up tomorrow at ten in the morning. We will go to the site, where I will have a small crew waiting to be briefed," Juan explained, knowing Mike would be agreeable.

"That sounds great, Juan. I'll see you in the morning. After the briefing, you can show me some of the sites, and we can get someone to show Mary and Sarah where the school is," Mike said, shaking Juan's hand.

"Yes, sir!" Juan replied as he turned and headed to the limousine. He opened the door to head out to the limousine then turned and waved to everyone. "Have a good evening! You'll find my wife's world-famous lasagna in the refrigerator. All you need to do is put the oven on 425 degrees and bake for fifteen minutes to warm it up. Enjoy!" Juan departed and closed the door as he left.

Mary went into the kitchen and found the lasagna and warmed it up in the oven. Mike scanned the house and its furnishings. Pleased with the layout and the design of his new home, he yelled out to

Mary, "This is really nice, honey! They thought of everything including towels and soap for the bathrooms."

Mary yelled back to Mike her response from the kitchen. "That's nice, dear. Dinner will be ready in about fifteen minutes." Sarah helped her mother set up the kitchen table for dinner.

Everyone was exhausted from the long trip. They sat down to enjoy a home-cooked meal. It was the best lasagna Mike and Mary had tasted in a long time. The pasta dish took a toll on them. They could barely stay awake long enough to get their pajamas on. Sarah fell asleep as soon as her head hit her pillow. Mike was too tired to bother with pajamas; he just dropped his pants and fell into bed.

Mary cleaned up the kitchen so she wouldn't have to do it in the morning and sat down at the bay window overlooking the neighborhood. It was about 8:30 p.m. The old-style streetlamps were illuminated and casting a pale-yellow hue over the nicely landscaped walkway. There was very little traffic on the street. A couple of late-night joggers ran by the house, waving as they passed. Mary felt very comfortable in her new home and drifted off to sleep in the chair.

CHAPTER 25

The Gift

Thanksgiving found Mary, Mike, and Sarah Evans enjoying dinner with Juan, Isabelle, and Hector Ramirez at their humble apartment in the Bronx. The two families had grown very close over the several months since their first meeting.

Hector and Sarah attended the same school for gifted children. They shared many of their classes and had become best of friends. As the adults gathered to discuss all that adult stuff, which tended to bore Sarah and Hector to death, they slipped out into the hall to make up a few games of their own.

Hector was a bit of a daredevil and loved to push the envelope with his stunts. He had always enjoyed being on the edge of danger. He liked to hop on his bike down the three flights of stairs to the ground floor. He also loved to scrape park benches and fence rails with the bottom of his skateboard.

Sarah was a lot more sensible and enjoyed having a good time but didn't really like to put herself in harm's way. She had talked to Hector several times to try and get him to settle down on some of his stunts, but Hector was pretty headstrong and just couldn't put himself in a safe position on this issue. He loved the rush of adrenaline in his brain.

Hector signed to Sarah, explaining that he could ride the length of the hallway on his hands. Sarah signed back to him, "I believe you! But I don't want to see that. Let's go downstairs and play in the park across the street." Hector smiled and acknowledged that he liked the idea then started off with a run and did a handstand on his skateboard, rolling quickly to the stairs.

Hector's skateboard slipped out from under him just as he reached the stairwell. He snapped his wrist as he tried to brace himself against the guardrail. He tumbled head over heels down the flight of stairs to the landing below. Sarah's face was pale with fear as she ran down the hall to see if Hector was still alive.

As she reached the end of the hall and looked down at Hector's body on the landing, she noticed he wasn't moving. She pounced down the steps two at a time to get to Hector. She kneeled beside him and slapped him on the shoulder. Hector wasn't moving at all. She pushed him over on his back to check his pulse and see if he was breathing.

Sarah knew immediately that he was in serious trouble. He had no pulse, and his head was rolling back and forth on the floor with no support from his broken neck. Sarah's eyes welled up with tears as she tried to yell for help in vain. She said a little prayer to herself and put her hands down on the chest of her good friend.

Her hands glowed with a blue iridescence as small white sparks shot across Hector's chest. Sarah's eyes widened with anticipation as Hector started breathing again. Then he slowly regained consciousness. He looked up at Sarah and smiled. He put his head back down and rested a few minutes as he tried to understand what just happened.

Sarah signed to him, "Are you okay?"

Hector checked himself out and then signed back, "Yes, I think so!" Hector noticed something very strange. He could hear sounds in the stair well and the hallway. He motioned to his ear and stood up, listening to the noises around him, the mumbling of people's voices in the apartments around them. His face reflected an expression of fear and excitement. It was as though he woke up in another world.

Then Sarah slapped him and pushed him away from her. "You scared me to death! Why did you do that?" Sarah signed.

"I don't know. I was showing off my new trick," Hector signed back. His lips formed would-be words. He heard himself utter sounds from his own mouth. "Oh my god! I can speak and hear sounds!"

Hector signed to Sarah. "Thank you, Sarah! I thought I was dead! How did you do that?"

Sarah looked at him and signed, "I don't exactly know how. I just did! I've been able to fix things for a long time. I just don't know how or why!"

"Does your mom and dad know you have this gift?" Hector inquired.

"I've never told them because they wouldn't understand. I'm afraid that they would treat me as someone special, or they might not let me go out and play with my friends if they knew," she signed with concern on her face.

"Yeah! You're right. They probably would freak out and lock you up somewhere until you got old. It's better you don't tell them. We'll keep it our secret until you decide you want to let them know," Hector signed and grunted to her with empathy. Sarah smiled and walked back up the stairs to the apartment. Dinner would be ready soon, and she'd had enough excitement for this visit.

Little did Sarah know that the excitement had just begun. As she turned the corner to head down the hall to the apartment, Sarah saw a hideous creature stand in the hall apparently waiting for her. The creature was black as tar with eyes red as burning coals. Its head was that of a jackal, and its long canine teeth were stained with blood. It had the appearance of a tar-soaked werewolf standing upright and staring directly at Sarah.

Sarah froze in her tracks; she had nowhere to run. Hector turned the corner right behind her and bumped into her as she froze. He looked to see what was in Sarah's way and saw the ugly creature. Hector jumped in front of Sarah and placed himself bravely between her and the beast. They creature was amused at his vain attempt to protect her and chuckled at the action.

The creature spoke to them saying, "You pathetic fool, stand aside and save yourself. I only want the girl. I can drop you with a sneeze, you panty waste." Hector and Sarah both heard and understood the creature as it spoke to them. They were both shocked that the creature could speak to them so clearly. Hector wadded up his hands in a fist and stood his ground against the beast that was five times his size.

The creature roared his displeasure at the boy that dared stand in his way. His breath was foul and caused Sarah and Hector to convulse violently. Hector shook it off and stood firm in the path of the creature, staring into its eyes, trying to figure a way to defeat the beast. The creature started growing impatiently with the boy and stretched out its right hand.

Hector's body was drawn toward the creature, and he fought against the grip to his throat. The creature laughed as the boy desperately struggled to get free. Hector was determined to defeat the beast and wanted to get closer to find a weakness, but he didn't want to get closer this way. Sarah was horribly disturbed with the creature's powers. She could feel her body tingling and didn't understand why.

Sarah took a firm stance in the hall, planting her feet shoulder width apart, and cast her arm out, mimicking the creature. Surprisingly, the creature lost his balance and fell backward onto the floor with a tremendous thud. Hector was free from the choking grip of the beast. Once free, he ran toward the monster and kicked him in the head several times as hard as he could. The beast roared more from anger than pain.

The beast bounded to his feet with a single push and stood eye to eye with Hector. Hector was so small in comparison to the creature that he had to lean forward to make eye contact with him. Hector saw an opportunity to put the beast down himself buy kicking him in the throat. As he shot his foot into the neck of the beast, it reacted by grabbing its throat and falling forward face first into the floor.

The creature roared again with displeasure again, releasing a horrible stench into the hall. Hector held his breath and continued landing kick after kick into the neck of the beast. Several times he landed a successful blow to the throat of the beast. The creature lay motionless on the floor, giving Hector the opportunity to dance victoriously around his limp body.

Sarah knew that he was only giving Hector a moment to feel victorious so the pain of the blow would be more shocking to him. After all, that's what the monster feeds on—fear and chaos. Sarah, unable to speak, could only watch in horror as the beast rose up behind the dancing figure and struck him from behind. The blow was so forceful that it sent his small body the full length of the hall into the wall behind Sarah. Hector's small frail, body stopped with a thud and dropped lifeless to the ground in a mangled heap.

Sarah turned instinctively to help her friend and companion, but as she turned toward Hector's limp body the creature moved in for the kill. She swung back toward the beast to try and minimize the impact of the blow. As she turned, the creature exposed the talons of his hand, which were easily no match for her. She could only watch as the beast rose high into the air to deliver the final blow.

Suddenly, the beast exploded into a thousand pieces and hovered in a circle in the air in front of her. Then the pieces of the creature began to spiral into a large vortex and were drawn into the floor of the hall. Ever so quickly the creature was gone, leaving Sarah alone in the hall with the lifeless body of her friend and brave protector.

Sarah turned to see Hector's body on the floor, and she was overcome with emotion and began to weep uncontrollably. She felt a gentle, comforting touch on her shoulder and turned slowly to see who it might be.

Standing beside her was the image of a man dressed like an Indian. Quietly he spoke to her. "Fear not, young one. Your friend is not gone yet! Touch him now as you did on the landing of the stairs, and he will be fine once again." Sarah complied with his instructions, placing her hands on Hector's mangled body.

A moment later, Hector began breathing again and sat up next to Sarah. "Is he dead? Did you see me kick his ass, Sarah? He won't be messing with you anymore!" he signed as though nothing had happened. Sarah just smiled and brushed his hair back out of his face.

Hector signed at her, "Who's the man in the loincloth?" He signed to her again, nodding his head toward the Indian figure next to her.

"You can see him?" she signed surprisingly to Hector.

"Who couldn't see a naked man standing in the hall with a loincloth over his thing?" Hector signed back, trying to vocalize his comment. Then he smiled at Sarah and gave her a hug and got to his feet. "Let's go get some turkey. It should be ready now." Hector motioned to Sarah and started to walk toward the apartment.

Geronimo put his hand on Sarah's shoulder and explained, "You're not finished yet! You have to touch his right temple with your index finger." Sarah looked up at the Indian with a puzzled look. As if he was anticipating her next question, he continued, "He can't have a memory of anything that happened, Sarah. It's not safe for him to know. They would hurt him for answers if he knew. Just touch his right temple. You'll understand later."

Sarah puts her hand on Hector's shoulder as he walked past her. He turned to see what she wanted, and she placed her index finger to his right temple. Hector brushed her finger away and signed, "What are you doing? It's time to go eat dinner. Come on!" As he opened his mouth while signing, he heard himself making noises and started to jump up and down with excitement.

Hector shared with Sarah that he could hear and speak then rushed down the hall to the apartment where you could hear the screams of joy and excitement as Hector shares his new skill with his parents. Sarah looked up at the Indian, a little sad, and turned to walk away. "Sarah, you are very special, and when the time is right, you'll know why things were this way. Have faith in the one who sent you here."

Then Geronimo disappeared into thin air. Sarah shrugged as though nothing could surprise her anymore and headed down the hall to the noisy apartment.

CHAPTER 26

Construction Falls Behind

Mike Evans called a meeting of all his field supervisors to discuss the schedule of completion. The project had fallen behind on its deadline, and it looked as though they weren't going to meet the first stage of completion. This was imperative if they were to secure the deadline bonus of 2.7 million dollars.

Mike called the meeting to order and went over the schedule deadlines laid out on the charts and explained the value of each one. When he completed the brief, he explained that at this time they were weeks behind schedule and that was going to cost them each their portion of the 2.7-million-dollar bonus.

Charlie Martin stood up and asked Mike what he meant. "What do you mean that we're going to miss our portion of the bonus? We were never told that we had a bonus on this project." Mike looked at the rest of his front-line supervisors, and they all had a dumbfounded look on their face. Mike explained the way he did business to them.

"Gentlemen! I don't know how the work bonus structure works in the state of New York. I only know one bonus structure—that is my bonus structure, which simply put means that when we reach our objectives on schedule, the corporation pays us a production bonus. In this case, it will be 2.7 million dollars if we complete the first building by Christmas."

Everyone in the room perked up in anticipation of some good news. Mike continued, "That bonus is paid equally to every workman on the project. It's paid out equally from me down to the cleanup crew, which means that when we hit the first bonus point on the contract, each of the 118 workers on this project will receive a bonus of $22,833 or something close to that. My math skills aren't as good as some of you, but that should be close."

The room fell deafly silent for a moment. Then Charlie jumped up and yelled out, "Let's hear it for the boss!" Cheers erupted in the room, and it sounded like an AMWAY convention for several minutes. Then Mike calmed everyone down and explained, "Gentlemen! Please, we only get the bonus if we meet the first deadline, and right now, we are two weeks off schedule. We don't have a lot of time to bring it around."

Charlie Marin stood up and walked over to Mike. He shook his hand and looked square into his eyes. "Sir! We will be on schedule at the turn of next week. I personally guarantee it!" Mike was pleased with the response and adjourned the meeting.

Sure enough, at the turn of the second week, the project was on schedule and had picked up significant momentum. Mike was not surprised. He knew that when the men found out that the bonus was theirs and not reserved for the senior staff, they would put their backs into it. Mike was the most popular project manager in New York City right now. Word of his pay plan had spread through the city like wildfire, and every construction worker in the city wanted a piece of the action.

During the next supervisor meeting, concerns were addressed pertaining to the morale of the men. Charlie Martin stood and addressed this issue to Mike. Charlie had become the unofficial spokesman for the group. "Mike, the men are concerned about their jobs. They have addressed these concerns to me and asked that I bring them up in this meeting." Mike looked at Charlie, very surprised to hear there was a problem with the men.

"What is their concern, Charlie?" Mike inquired.

Charlie looked around the room for approving nods from all the supervisors there. Once received, he continued, "Well, the men are very concerned about their jobs, Mr. Evans. They are under the impression that they will be replaced by more experienced labor now that word is out that the deadline bonus is paid to the workers."

Mike couldn't hold back his laughter. "I'm sorry! I'm not laughing at their concern—I'm just very overwhelmed by it. Why do the men think that I would replace them with other people in the city?" Mike looked at the faces of all the supervisors in the room then turned to Charlie for his response.

"Well, the men felt that the announcement of the shared bonus was a corporate move to attract more experienced labor on the job. Once enough people responded with interest to work on the project, they would then be replaced, and the bonus would go to them." Charlie finished his remarks then sat down with the rest of the supervisors, waiting for Mike's response.

Mike remained silent for a moment, adding to the tension. Then softly but firmly, he responded, "I've been working this way since I've been in the construction business. I believe in my workers, and my workers are the ones that I chose for this job. That includes each of you. I can't speak for anyone else in this business. I can only speak to you pertaining to myself."

Mike took a seat at the table and extended his hands to Charlie on his right and Evan Morris on his left. He indicated with a nod for everyone to join hands at the table. Slowly all hands were connected around the table. "Each of you took a chance on me when I arrived here in New York City. I was the outsider who slipped in the back door. Yet each of you came forward against the instruction of your friends and relatives and placed your trust in me."

Mike paused for a moment and took a few deep breaths. "My name is Mike Evans, and I am the project manager for this construction group. I want you all to know without a doubt that my selection is final. Whether we succeed or fail, this is the team that we will succeed or fail with. (He held up the hands of each supervisor at the table.) I assure you, gentlemen, that with the same faith that you place in me, I place in you to complete this job. I won't replace any of you on this project, nor will I ever replace any of the men we have chosen. Unless it is their will to do so."

Mike released their hands and stood up. He advised them to take that message to each of the workers. "Now I want each of you to take that message exactly the way I've delivered it to you and share it with our crew." Mike terminated the meeting and walked out of the room emotionally exhausted.

Charlie turned to the rest of the men in the room. "Each of us should be ashamed for listening to rumor and gossip. We need to make a pact here and now that we will never doubt the integrity of Mr. Mike Evans ever again. All in favor nod your head." Everyone in the room complied with Charlie's request and each left the room in somber silence.

Charlie, Evan, and the rest of the supervisors took the message back to the workers just the way Mike wanted it. The following week, when rumors started again across the workforce, they were put down immediately by the workers and supervisors. From that day forward, there were no disgruntled employees on the team. Everyone focused intently on completing the project on time, and morale skyrocketed.

CHAPTER 27

Sarah's Song

Sarah slipped into the back door of St. Paul's Cathedral and let go a sigh of relief. Finally, after all she had been through to get there, she felt total peace overcome her. She smiled as she walked slowly to the dressing area, where her friends were getting ready for the show. She walked into the dressing area with her clothes torn and dirt smudges on her face.

She sat at one of the table's set up for makeup and began cleaning her face. Mrs. Westermann walked in to see Sarah at the dressing table by herself. She walked over and grabbed a brush and helped her with her hair, offering Sarah a most pleasing smile and confident nod.

The Christmas show had already begun, and most of the other students were on stage performing their hearts out. Mrs. Westermann knew that Sarah wouldn't be allowed to perform tonight but took great joy in helping her get ready to stand with the rest of the students at the end for the final bow.

Sarah cleaned up nicely and selected a beautiful all-white silk dress with no fancy designs or fringe, a simple white dress, which on her looked beautiful. Her hair was put up in a bun on top of her head, and Mrs. Westermann took a small strand of her hair and curled it down each side of her face. She looked like an angel, Mrs. Westermann thought to herself when they were done.

Mike and Mary finally arrived with the help of their new friend James. They entered the front of the cathedral and were awestruck by the turnout for this production. The crowd that filled St. Paul's cathedral was twice the size of Carnegie Hall. Mike was still upset that the program director Mrs. Grissom decided again to have Sarah sign the production for those in attendance that were deaf.

He trusted what James and Sarah explained to him earlier. He wanted so much to hear a sound come from the mouth of his lovely daughter that he would believe anything that offered him the chance to hear that. James looked at the stage scanning for Sarah, but there was no sign of her.

The echo of siren horns and screams still nested in Mike and Mary's ears. They tried to shake it but were unsuccessful. Finally, James saw Sarah standing next to a lady to the right of the stage. She looked so radiant that James began to tear up. He pointed her out to Mike and Mary, and they too became softhearted toward her resolve. Deep inside, Mike and Mary were very proud of their little Sarah for her tenaciousness.

The children sang and signed their hearts out on this production, and everyone was moved by their achievement. Mrs. Grissom noticed Sarah standing on the side of the stage and motioned for her to come over and stand next to her. Sarah timidly walked over and stood next to the director as she moved to the finale.

Then the house lights dimmed, and little Marsha Pringle walked out onto the stage to sing silent night. As she opened her mouth to sing, nothing came out only a wheezing sound. The audience gasped, and the music stopped. Marsha rubbed her throat and nodded to start again.

Mrs. Grissom was relieved and nodded to the orchestra to start the lead in again. When it got to the first note that Marsha was supposed to sing, she wheezed and then ran off the stage to the wings, where Mrs. Westermann gave her a hug and tried to talk to her and see what was wrong.

After a moment Mrs. Westermann looked over at Mrs. Grissom and shook her head, indicating that Marsha wouldn't come back on stage. Mrs. Grissom turned to announce to the audience that the show was over. As she turned, she noticed that Sarah had boldly walked to the stage and picked up the microphone.

The audience applauded their encouragement for little Sarah. She stood quietly and confidently in front of the crowd and nodded to Mrs. Grissom to begin. Mrs. Grissom nervously nodded to the orchestra to start the lead in again. The audience became silent, and the orchestra began to play.

Everyone in the orchestra, all the children and adults on stage and most of the people in the audience, knew that Sarah was deaf and mute and had been so since birth. They were surprised to see her holding the microphone as though she were going to sing the song.

As the orchestra reached the part where the little girl was to start singing, they heard a loud noise outside the cathedral. The orchestra stopped playing again. Sarah looked to the back of the church to see the Knights of Columbus enter, dressed in their fine traditional dress. With their swords drawn and held up in front of them, they marched into the cathedral, making a path for the Pope.

The Knights spread through the cathedral in a column of twos and dividing the aisles. They raised their swords over their heads and crossed them as the Pope walked in and headed to the front of the cathedral. The Knights of Columbus were assigned the personal responsibility of protecting the Pope and all papal authority.

They were dressed in all-black tuxedos with beautiful, feathered head covers of man colors and with a flowing black cape with a cardinal red lining. Their shiny sword blades glistened with holy oil. The Pope traveled down the aisle carved out by the Knights to the front of the cathedral, offering a blessing to all in attendance.

At the front of the cathedral, he shook hands with Mrs. Grissom and motioned her to continue. The Knights sheathed their swords and faced center stage. Mrs. Grissom turned back to the orchestra and nodded to the orchestra and looked back to Sarah and offered a confident smile.

Sarah opened her mouth, and the words of "Silent Night" poured from her lips as an angelic beacon to the world. Her voice was pure and soothing to all. Everyone gasped in the audience when they heard Sarah sing like an angel. Her voice carried through the cathedral and out into the streets of New York City.

Over the noise of the city, you could hear her angelic voice, and as people heard it, they stopped in their tracks and offered a somber thanksgiving for such a beautiful sound.

People looked up to the sky over St. Paul's Cathedral and saw hordes of angels descending to the cathedral. They looked on with fear and amazement, not understanding what was happening. Those who witnessed the marvelous event were overcome with peace and love for their fellow man.

In the cathedral, Pope John Paul stood and raised his hands to heaven, inviting the angels to join them in the audience. The audience looked up to the ceiling of the great cathedral and witnessed more angels than they could count filling the rafters and the arches sitting and listening to the beautiful voice of Sarah's song.

Pope John Paul's eyes filled with tears as he witnessed the miraculous event unfolding before everyone and offered praising gestures to the Lord on behalf of the flock that sat in witness.

Sarah's voice traveled through the streets of New York and into every ear; none were spared as her voice reached each soul. Violence in the city disappeared and was replaced with love and respect for each other. The gangs that were preparing to defend their turf from rival gangs dropped their weapons into the streets and invited their rivals to sit with them.

Every city in the United States heard the voice of Sarah's Song and stopped in their tracks to offer thanks to their belief. They joined hands in the streets to welcome in the new Christmas celebration and to offer a gesture of friendship to all they encountered.

An army of angels carried Sarah's Song to the ends of the earth, ensuring that every ear was filled with the blessing of God. Warships out to sea, were called back immediately. Missiles around the world were disarmed. The Jews and Palestinians dropped their guns in the streets and embraced each other in fellowship.

The terrorists that had been programmed by Dr. Swanson and wearing bombs in major cities across the United States, Canada, United Kingdom, and other countries friendly to them turned themselves into local authorities and disarmed the bombs. The song had softened their hearts and had dissolved the triggering mechanism implanted in their brains.

Their selfless act of peace was well-received by authorities as they approached the local police stations to turn themselves in. The police welcomed them with love and fellowship and helped them remove the objects of destruction. Then they invited them to share Christmas supper with them. No charges were filed against them, and they were released back to their families.

Sarah's song was a two-edged sword. Those who heard the voice of Sarah's song and received God's message of peace and love, their hearts were softened, and they were cured of the evil that had overcome them. Those who were totally consumed with hate and hosted hearts of stone, when they heard the voice of Sarah's song, their hearts exploded in their chest, and they dropped dead where they stood.

The counsel of evil who had conspired to take over the world through violence and force all people to surrender to their demands fell silent with no more than a small thud. Their chests burst as they tried to execute their plan of mass destruction. Each sat in horror as one by one their chests exploded, leaving them lifeless on the meeting room floor.

When the final leader hit the floor, the soldiers of souls arrived to claim their prize. As their spirits departed their bodies, they were greeted by Satan's spawn and screamed to God to bring them out of damnation. But as their hearts were turned to stone against his will, he had no ear for their cries of mercy. It was too little too late for them.

Six hundred and sixty-six thousand, six hundred and sixty-six fell victim to Sarah's song. This was the number of pure evil in the world at the time of God's miracle, just as it was foretold in the thirteenth scroll that was discovered by Tim Wilford and his colleague Haseeb Jaffar.

In the deserts around the world, Satan's army of evil was massing to confront God's army of angels and push them back from earth. After the angels had carried Sarah's song to every ear on earth, they divided their forces and attacked the armies of Satan and forced them back into the pit of eternal fire and brimstone, casting them into the bottomless pit and locking them in with the seal of the thirteenth scroll. All this took place in a single hour after Sarah sang "Silent Night" in St. Paul's Cathedral. It was as though time had stood still until the battle was over and Sarah's song was presented to the world.

Back at the cathedral, Sarah finished singing her song. Not a sound could be heard in the entire city. The world stood still in silence, reflecting on the great miracle they had just witnessed. There was no reason to clap for Sarah, only a sincere effort to offer quiet reflection on the peace that was restored to a world that was so close to destruction.

After ten minutes of silence, Pope John Paul approached Sarah Ann Victoria Evans (SAVE) and kneeled before her, and blessed her as a true saint and the bearer of God's word to the world. Sarah reached out her hand to the Pope and lifted him up from his knees. Then she knelt before the Pope and respectfully bowed before him as the keeper of the keys to the Kingdom of God and dually elected authority of God's Church who sat on the throne of Peter.

Mary and Mike Evans were called forward by the Pope to stand beside their daughter. The Angels rejoiced and sang songs of praise to God Almighty and raised their hands and voices to the heavens. The people joined in and sang with the angels, and even the Pope, who was never known for his ability to carry a single note, joined in with his hands raised to God. The Knights of Columbus raised their swords in to salute to God the Father, Mary our immaculate Mother, and Jesus the Son of God, and redeemer of man.

Mike and Mary hugged and kissed Sarah. Mike picked her up and placed her on his shoulders so she could raise her hands higher than anyone else could. As Sarah raised her hands to the heavens, all the angels flew down to touch her hand one after another. With each touch, Sarah received another blessing from heaven.

For twelve long years, Sarah was deaf and dumb to the world around her, and now in the time it took for her to sing one song, she had received her hearing and voice and been blessed by every angel in heaven. She was about to begin a wonderful new life on earth as the child who brought peace and harmony back to a world that was on the edge of destruction.

Mike, Mary, and Sarah Evans were about to receive the most wonderful gift of all. In God's mercy, he raised them up with the angels and brought them into his grace for all time. God knew that the world wasn't through with its growth and the human challenges that came from it. He decided to spare them from what was about to unfold on the planet.

Mike, Mary, and Sarah Evans were quickened in the blink of an eye, and their transformation was witnessed by all those present in the cathedral, including the Pope. Their physical appearance changed to a bright, glowing light that presented a warm and comforting feeling to everyone who witnessed it.

And with the angels, they ascended to heaven to sit in the host of angels in the presence of God the Father.

The End

About the Author

Authors Kim Richard Smith (author of *The Dark Figure*) and Kevin James Smith were born in Panama City, Florida, five minutes apart. They wrote *Sarah's Song* after collaborating for a year. They reached

out to Dee Arianne Rockwood for her commentary, which was significant in the completion of this book. Kim served in the US Navy during Vietnam and did three tours. Kevin served with the US Air Force, working underground in a vault. Dee worked with the CIA and is a black belt karate champion. Kim and Kevin moved all over the country, moving from base to base as their father was an Air Force ace pilot. Kim now lives in Waverly, Ohio, and Kevin lives in Gilbert, Arizona.